ADVANCED CONTROL WITH
MATLAB AND SIMULINK

ADVANCED CONTROL WITH MATLAB AND SIMULINK

edited by
Jerzy Mościński
Silesian Technical University, Poland

Zbigniew Ogonowski
Silesian Technical University, Poland

ELLIS HORWOOD
LONDON NEW YORK TORONTO SYDNEY TOKYO SINGAPORE
MADRID MEXICO CITY MUNICH

First published 1995 by
Prentice Hall Europe
Campus 400, Marylands Avenue
Hemel Hempstead
Hertfordshire, HP2 7EZ
A division of
Simon & Schuster International Group

Printed and bound in Great Britain by
Ashford Colour Press Ltd, Gosport, Hampshire

Library of Congress Cataloging-in-Publication Data

Available from the publisher

British Library Cataloguing in Publication Data

A catalogue record for this book is available
from the British Library

ISBN 0-13-309667-X

4 5 99

Although every effort has been made to ensure that the MATLAB scripts and functions provided are error free, neither the authors nor the publisher shall be held responsible or liable for any damages resulting in connection with or arising from the use of any of the functions or scripts in this book.

MATLAB is a registered trademark of The Mathworks, Inc., 24, Prime Park Way, Natick, MA, USA, 01760-1500. Tel +(508) 653-1415, FAX +(508) 653-2997, e-mail: info@mathworks.com.

Other trademarks mentioned in the text include; Control-C, a registered trademark of SCT, Inc., Derive, a registered trademark of Soft Warehouse, Inc., IBM PC, a trademark of International Business Machines, Inc., Macintosh, a trademark of Apple Computers, Inc., Maple, a registered trademark of Waterloo Maple Software, Mathematica, a trademark of Wolfram Research, Inc., MS Windows, a trademark of Microsoft, Inc., NAg, a registered trademark of NAG Ltd., Sparc and Sun, trademarks of Sun Microsystems, Inc., X Window, a trademark of Massachusetts Institute of Technology.

List of contributors

Ewa Bielińska
Silesian Technical University, Gliwice, Poland

Mieczysław A. Brdyś
University of Birmingham, United Kingdom

Jarosław Figwer
Silesian Technical University, Gliwice, Poland

Jerzy Kasprzyk
Silesian Technical University, Gliwice, Poland

Ulrich Keuchel
Ruhr Universität, Bochum, Germany

Jerzy Mościński
Silesian Technical University, Gliwice, Poland

John P. Norton
University of Birmingham, United Kingdom

Zbigniew Ogonowski
Silesian Technical University, Gliwice, Poland

Contents

Foreword **xiii**

1 Basic system dynamics **1**

1.1 Introduction 1
1.2 System modelling 1
1.2.1 System models and the concept of state 1
1.2.2 State-space representation of linear systems 3
1.2.3 State transformations 12
1.2.4 Transfer function 13
1.3 Solution of the state equation 15
1.3.1 The homogenous solution – transition matrix 15
1.3.2 Calculation of the transition matrix 16
1.4 System stability 18
1.4.1 Stability concepts and definitions 18
1.4.2 Routh–Hurwitz criterion 19
1.4.3 Lyapunov's method 20
1.5 Controllability and observability 21
1.5.1 Controllability 21
1.5.2 Observability 23
1.6 Feedback control and state estimation 25
1.6.1 Pole placement 25
1.6.2 Control loop with state-feedback controller and observer 26
1.6.3 Synthesis of quadratic optimal control loops 28
1.7 Conclusions 33
1.8 Laboratory exercises 34
1.8.1 MATLAB software tools applied in laboratory course 1 34
1.8.2 Continuous-time systems: transfer-function and state-space representation 34
1.8.3 Discrete-time systems. 39
1.8.4 System stability 40
1.8.5 A design technique based on the elementary methods of stability analysis 41
1.8.6 System controllability and observability 42
1.8.7 Controllability and observability in feedback system design 43
1.8.8 Using an S-function for simulation of complex systems 45

1.9 References 51

2 Control system design **53**
2.1 Introduction 53
2.2 SISO plant representation 53
2.2.1 Transfer function and frequency response 53
2.2.2 Uncertainty 54
2.2.3 Signal and system measures 56
2.3 Feedback system 58
2.3.1 Basic concepts 58
2.3.2 Stability 62
2.3.3 Root locus 63
2.3.4 Robust stability and performance 63
2.4 Classical design principles 66
2.4.1 Stability margin 66
2.4.2 Sensitivity and complementary sensitivity 67
2.4.3 Compensation 69
2.4.4 PID control 70
2.4.5 Pole placement 72
2.5 Multivariable plant 73
2.5.1 Representation and uncertainty 73
2.5.2 Interaction 74
2.5.3 Signal and system measures 75
2.6 Multivariable feedback systems 76
2.6.1 Performance 76
2.6.2 Stability 78
2.6.3 Robust stability 79
2.7 Multivariable control design 80
2.7.1 Control structure design 80
2.7.2 Diagonal control 81
2.7.3 The characteristic locus method 82
2.7.4 The Nyquist array method 83
2.8 Linear/quadratic control 83
2.8.1 Linear/quadratic control problem 83
2.8.2 Linear/quadratic Gaussian control 84
2.8.3 Loop transfer recovery 87
2.9 H_∞ control 88
2.9.1 Two-port control problem representation 88
2.9.2 H_∞ control problem 89
2.10 Laboratory exercises 90
2.10.1 MATLAB software tools applied in laboratory course 2 90
2.10.2 SISO plant representation 91
2.10.3 Feedback systems 93
2.10.4 Classical system design 96
2.10.5 Multivariable control systems 98
2.10.6 Linear/quadratic control 102

2.10.7 H_∞ control 104
2.11 References 106

3 Process identification **108**
3.1 Introduction 108
3.2 Aims of identification 108
3.2.1 Insight 108
3.2.2 Prediction 109
3.2.3 State estimation 109
3.2.4 Simulation 110
3.2.5 Fault detection 110
3.3 Types of models 111
3.3.1 Dynamics 111
3.3.2 Linearity 111
3.3.3 Discrete- and continuous-time models 112
3.3.4 Time-invariance 112
3.3.5 Other properties of the model 113
3.4 Model types and test signals 113
3.4.1 Non-parametric SISO time-domain models 113
3.4.2 Non-parametric SISO frequency-domain models 117
3.4.3 Parametric SISO time-domain models 118
3.4.4 SISO transform models 118
3.4.5 MIMO models 121
3.5 Identification algorithms 121
3.5.1 Batch ordinary least-squares estimation 121
3.5.2 Batch weighted-least-squares estimation 123
3.5.3 Recursive ordinary-least-squares estimation 123
3.5.4 Recursive weighted-least-squares estimation 125
3.5.5 Information updating 125
3.5.6 Algorithms to avoid bias due to correlation between regressors and noise 125
3.5.7 Identification of time-varying systems 127
3.6 Model-structure selection 128
3.7 Model validation 129
3.8 Experiment design 130
3.9 Topics not covered 131
3.10 Laboratory exercises 131
3.10.1 MATLAB software tools in laboratory course 3 131
3.10.2 Simulation and prediction of SISO transform models 132
3.10.3 Model types and test signals 133
3.10.4 Identification algorithms 140
3.10.5 Model structure selection and model validation 146
3.11 References 148

4 Neural networks in identification and control **150**
4.1 Introduction 150
4.1.1 Origins of connectionist research 150
4.1.2 Neural networks and control 150
4.2 Network architecture 151
4.2.1 Neurons 152
4.2.2 Statical multi-layer feed-forward networks 153
4.2.3 Feedback (recurrent) networks 156
4.3 Learning in statical networks 158
4.3.1 Learning rules for a neuron 158
4.3.2 Delta learning rule for multilayer feed-forward networks and back-propagation training algorithm 162
4.4 Learning in dynamical networks 165
4.4.1 Back-propagation through time 165
4.4.2 General dynamical back-propagation 166
4.5 Identification 167
4.5.1 Forward modelling 167
4.5.2 Inverse modelling 170
4.6 Control 172
4.6.1 Model reference control 172
4.6.2 Internal model control (IMC) 173
4.6.3 Predictive control 173
4.6.4 Self-learning controller 174
4.6.5 Non-linear self-tuning adaptive control 176
4.7 Conclusions 179
4.8 Laboratory exercises 180
4.8.1 MATLAB software tools applied in laboratory course 4 180
4.8.2 Single neuron and learning 181
4.8.3 Off-line neural model identification 186
4.8.4 Recursive identification 196
4.8.5 Neural networks for control 202
4.9 References 207

5 Adaptive control **210**
5.1 Introduction 210
5.1.1 Adaptive control schemes 210
5.1.2 Plant model 213
5.2 Control algorithms suitable for adaptive systems 215
5.2.1 Minimum-variance control algorithms 215
5.2.2 Pole/zero-placement algorithms 216
5.2.3 Simple self-tuning control algorithms 217
5.2.4 MAC control algorithm 218
5.2.5 GPC algorithm 220
5.2.6 Adaptive PID control with non-parametric identification 221
5.3 Identification in adaptive systems 221
5.3.1 Recursive least squares 222

5.3.2 Recursive least mean squares 225
5.4 Identification model structure 226
5.4.1 Minimum phase plant 226
5.4.2 Non-minimum phase plant 226
5.5 Estimation of non-stationary plant parameters 228
5.6 Multivariable minimum-variance adaptive control 228
5.7 Multivariable predictive adaptive control 231
5.8 Multivariable system identification 233
5.9 Performance measures for adaptive control algorithms 235
5.10 Conclusions 238
5.11 Laboratory exercises 238
5.11.1 MATLAB software tools applied in laboratory course 5 238
5.11.2 Adaptive control 238
5.12 References 250

Foreword

The courses covered by this book were presented during the third Summer School organized under TEMPUS project JEP 0962-91/93 "Education in Control Systems and Information Technology". The JEP initiated various activities within several participating institutions from the European Community, Poland and the Slovak Republic. In the first year of the project, the emphasis was on tools for control system design. The second year was devoted mainly to real-time control and image processing. Advanced control concepts and tools were the key issue of the final year of the project.

The third Summer School took place in the Institute of Automatic Control of the Silesian Technical University, Gliwice, Poland, in June 1993. The course was divided into five parts: *Basic system dynamics* prepared by Ruhr Universität Bochum, Germany, *Control system simulation* prepared by Cambridge Control, Cambridge, UK, *Process identification* prepared by the University of Birmingham, UK, *Neural networks in identification and control* prepared by the University of Birmingham, UK, and *Adaptive control* prepared by the Technical University of Denmark. All parts were supported with sets of laboratory exercises prepared by the Silesian Technical University. Now the results of that international effort are published in book form. The presentation of the state of the art of advanced control concepts and tools, together with ready–to–use laboratory exercises prepared in the MATLAB®/SIMULINK® environment, should be useful both in teaching and in engineering practice.

The organization of the book corresponds to the Summer School contents. In the five parts mentioned above, laboratory exercises followed lectures. Dr Keuchel is the author of the lecture part of Chapter 1, Dr Bielińska is the author of the laboratory exercises part of this chapter. Dr Figwer prepared both lecture and laboratory exercises parts in Chapter 2. Professor Norton is the author of the lecture part of Chapter 3, Dr Kasprzyk wrote the laboratory exercises part of it. Professor Brdyś prepared the lecture part of Chapter 4, Dr Ogonowski is the author of the laboratory exercises part of this chapter as well as the laboratory exercises part in Chapter 5. Finally, Dr Mościński prepared the lecture part of Chapter 5.

Laboratory exercises prepared for the third Summer School and quoted in this book included some dedicated software written specifically for them. This software, in the form of MATLAB/SIMULINK m-files, is available freely from the authors. The preferred way of doing this is by *anonymous ftp*. The reader who is interested in obtaining these m-files should `ftp` to `ftp.appia.polsl.gliwice.pl` (or `157.158.13.7`), give `anonymous` as user name and e-mail address as password, and look for the software in directory `/pub/appia/AdvancedControl`. The m-files can also be found through The MathWorks anonymous ftp site at `ftp.mathworks.com` in `/pub/books/moscinski`.

The editors would like to take this opportunity to thank all authors of the contributions to the book, and also would like to express gratitude to all who contributed to the success of the Summer School. It is hoped that the efforts of all the people involved will benefit the academic and industrial community through this book.

Jerzy Mościński
Zbigniew Ogonowski

March, 1995

1

Basic system dynamics

1.1 INTRODUCTION

The behaviour of real dynamical systems is often too complex for complete mathematical analysis. The engineer is obliged to use approximate descriptions which deliver sufficiently accurate results at acceptable cost. Substantial simplifications of the mathematical description, analysis and design of control systems can be obtained by the following assumptions:

- the relation between input and output variables of the process is linear,
- the process does not change its properties during the observation period,
- all process variables have a continuous-time course and are measured continuously.

Processes which exhibit these properties, at least approximately, can be described using quite simple mathematical tools. The theory for linear time-invariant systems is well established and is standard in all basic courses on automatic control.

This chapter will serve as a tutorial on some fundamentals of linear system theory. Concepts of system modelling, state equations, stability, controllability and observability, including basic design approaches for multivariable, continuous-time and discrete-time systems, will be presented as a basis for the following more specific chapters.

We start with an introductory section on differential equations and state-space descriptions for linear multivariable systems. Special forms of state-space models and the concept of similarity transformation will be described. In addition, input–output descriptions for linear systems by means of the transfer-function matrix will be briefly discussed followed by the state-space representation consideration in time and frequency domains. Stability definitions and stability criteria, observability and controllability concepts, and design methods for state feedback conclude the chapter.

1.2 SYSTEM MODELLING

1.2.1 System models and the concept of state

The signals generated by dynamical systems and the systems themselves can be described by differential equations and difference equations.

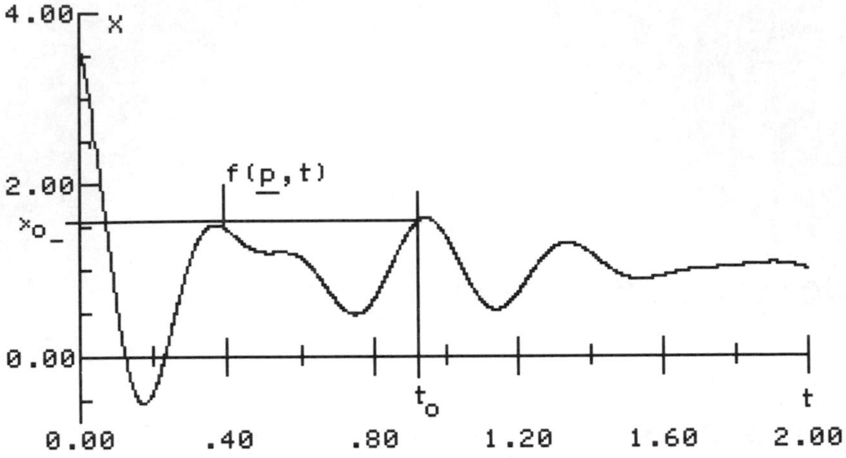

Figure 1.1. Deterministic continuous-time signal

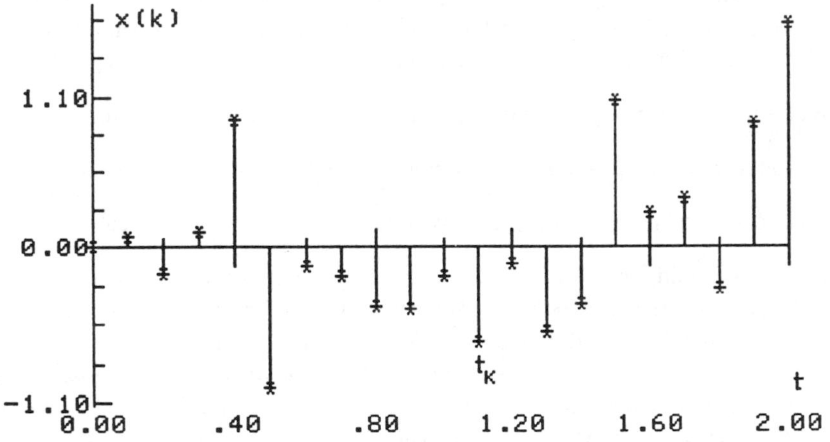

Figure 1.2. Deterministic discrete-time signal

In the continuous-time, deterministic case a signal $x(t) = f(p, t)$ (see Fig. 1.1) can be described by a homogeneous differential equation

$$x^{(n)} = F(x, x^{(1)}, x^{(2)}, \ldots, x^{(n-1)}, p), \tag{1.1}$$

with initial conditions

$$x(0), x^{(1)}(0), x^{(2)}(0), \ldots, x^{(n-1)}(0), \tag{1.2}$$

where p denotes a vector containing the parameters. A very simple example is

$$x(t) = A \sin(\omega t + \varphi). \tag{1.3}$$

In this case the parameterization is

$$p = \begin{bmatrix} A & \omega & \varphi \end{bmatrix}^T. \tag{1.4}$$

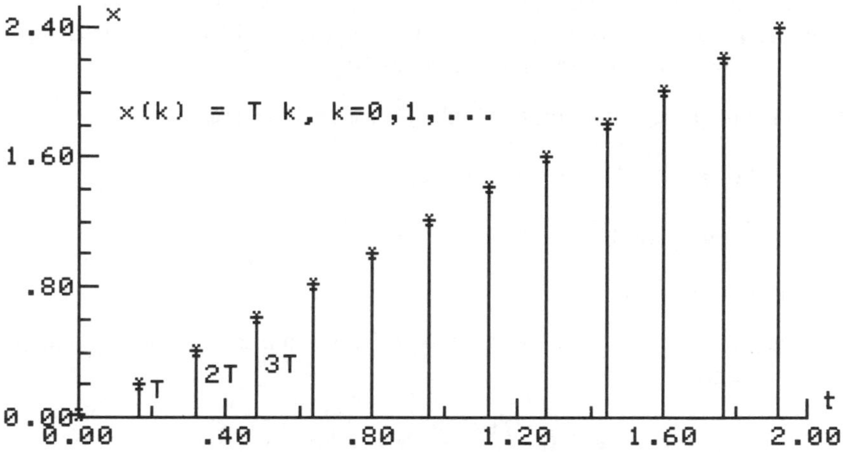

Figure 1.3. Discrete-time ramp signal

Alternatively, the associated differential equation is

$$\ddot{x} + \omega^2 = 0, \tag{1.5}$$

with initial conditions

$$\dot{x}(0) = f(A, \varphi, \omega) = \omega A \cos(\varphi), \quad x(0) = g(A, \varphi) = A \sin(\varphi). \tag{1.6}$$

In the discrete-time deterministic case a sampled signal $x(k) = f(p, k)$, $k = 0, 1, \ldots$ (see Fig. 1.2) can be described by a homogeneous difference equation

$$\begin{aligned} x(k) &= F(x(k-1), x(k-2), \ldots, x(k-n), p) \\ &\qquad x(0), x(1), x(2), \ldots, x(n-1). \end{aligned} \tag{1.7}$$

Here, the initial values are replaced by shifted sampled values of the signal. An example of a difference equation, describing the ramp signal depicted in Fig. 1.3, is

$$x(k) = x(k-1) + T. \tag{1.8}$$

1.2.2 State-space representation of linear systems

State and output equations

A differential equation of nth order can be rewritten as a system of n differential equations of first order and several additional purely algebraic identities. This is possible for non-linear differential equations as well as for linear differential equations. Generally:

$$\begin{aligned} \frac{d\boldsymbol{x}(t)}{dt} &= \boldsymbol{f}[\boldsymbol{x}(t), \boldsymbol{u}(t)], \qquad \boldsymbol{x}(0) = \boldsymbol{x}_0 \\ \boldsymbol{y}(t) &= \boldsymbol{g}[\boldsymbol{x}(t), \boldsymbol{u}(t)]. \end{aligned} \tag{1.9}$$

Many control algorithms are of the linear type (PID, state-variable feedback), whereas models of the controlled systems (chemical plants, robot manipulators) are in most situations related to non-linear dynamical equations. To prepare models for analysis and synthesis of control systems we have to linearize the system equations (1.9) for a specific operating point. The procedure for doing this involves three steps:

1. Define a nominal operating point $(\boldsymbol{x}^0, \boldsymbol{u}^0, \boldsymbol{y}^0)$ which is governed by

$$\dot{\boldsymbol{x}}^0 = \boldsymbol{f}(\boldsymbol{x}^0, \boldsymbol{u}^0), \quad \dot{\boldsymbol{y}}^0 = \boldsymbol{g}(\boldsymbol{x}^0, \boldsymbol{u}^0)$$

2. Consider an input perturbation $\boldsymbol{u} \to \boldsymbol{u}^0 + \boldsymbol{\delta u}$

3. Linearize around $(\boldsymbol{x}^0, \boldsymbol{u}^0, \boldsymbol{y}^0)$

The linearization is usually done by expanding the non-linear differential equation (1.9) using Taylor's series

$$\begin{aligned}\dot{\boldsymbol{x}}^0 + \boldsymbol{\delta}\dot{\boldsymbol{x}} &= \boldsymbol{f}(\boldsymbol{x}^0, \boldsymbol{u}^0) + \frac{\partial \boldsymbol{f}(\boldsymbol{x}^0, \boldsymbol{u}^0)}{\partial \boldsymbol{x}^T}\boldsymbol{\delta x} + \boldsymbol{\delta x}^T \frac{\partial^2 \boldsymbol{f}(\boldsymbol{x}^0, \boldsymbol{u}^0)}{\partial \boldsymbol{x}^2}\boldsymbol{\delta x} + \cdots &(1.10)\\ &+ \frac{\partial \boldsymbol{f}(\boldsymbol{x}^0, \boldsymbol{u}^0)}{\partial \boldsymbol{u}^T}\boldsymbol{\delta u} + \boldsymbol{\delta u}^T \frac{\partial^2 \boldsymbol{f}(\boldsymbol{x}^0, \boldsymbol{u}^0)}{\partial \boldsymbol{u}^2}\boldsymbol{\delta u} + \cdots &(1.11)\end{aligned}$$

Neglecting higher-order terms and taking into account $\dot{\boldsymbol{x}}^0 = \boldsymbol{f}(\boldsymbol{x}^0, \boldsymbol{u}^0)$,

$$\boldsymbol{\delta}\dot{\boldsymbol{x}} = \frac{\partial \boldsymbol{f}(\boldsymbol{x}^0, \boldsymbol{u}^0)}{\partial \boldsymbol{x}^T}|_{\boldsymbol{x}=\boldsymbol{x}^0}\boldsymbol{\delta x} + \frac{\partial \boldsymbol{f}(\boldsymbol{x}^0, \boldsymbol{u}^0)}{\partial \boldsymbol{u}^T}|_{\boldsymbol{u}=\boldsymbol{u}^0}\boldsymbol{\delta u} \tag{1.12}$$

for small deviations from the nominal operating point. The derivative of a vector with respect to a transposed vector is the Jacobian matrix, so Eq. (1.12) can also be written component-wise as

$$\begin{aligned}\begin{bmatrix}\delta\dot{x}_1\\ \vdots \\ \delta\dot{x}_n\end{bmatrix} &= \begin{bmatrix}\frac{\partial f_1}{\partial x_1} & \cdots & \frac{\partial f_1}{\partial x_n}\\ \vdots & \ddots & \vdots\\ \frac{\partial f_n}{\partial x_1} & \cdots & \frac{\partial f_n}{\partial x_n}\end{bmatrix}\begin{bmatrix}\delta x_1\\ \vdots \\ \delta x_n\end{bmatrix} &(1.13)\\ &+ \begin{bmatrix}\frac{\partial f_1}{\partial u_1} & \cdots & \frac{\partial f_1}{\partial u_m}\\ \vdots & \ddots & \vdots\\ \frac{\partial f_n}{\partial u_1} & \cdots & \frac{\partial f_n}{\partial u_m}\end{bmatrix}\begin{bmatrix}\delta u_1\\ \vdots \\ \delta u_m\end{bmatrix}. &(1.14)\end{aligned}$$

The same series expansion for the output equation gives

$$\boldsymbol{\delta y} = \frac{\partial \boldsymbol{g}(\boldsymbol{x}^0, \boldsymbol{u}^0)}{\partial \boldsymbol{x}^T}|_{\boldsymbol{x}=\boldsymbol{x}^0}\boldsymbol{\delta x} + \frac{\partial \boldsymbol{g}(\boldsymbol{x}^0, \boldsymbol{u}^0)}{\partial \boldsymbol{u}^T}|_{\boldsymbol{u}=\boldsymbol{u}^0}\boldsymbol{\delta u}, \tag{1.15}$$

so that the linear state-space system will be described by

$$\begin{aligned}\boldsymbol{\delta}\dot{\boldsymbol{x}} &= \boldsymbol{A}\boldsymbol{\delta x} + \boldsymbol{B}\boldsymbol{\delta u} &(1.16)\\ \boldsymbol{\delta y} &= \boldsymbol{C}\boldsymbol{\delta x} + \boldsymbol{D}\boldsymbol{\delta u}. &(1.17)\end{aligned}$$

If only linear systems are considered, it is convenient to replace Eq. (1.16) by

$$\begin{aligned}\frac{d\boldsymbol{x}(t)}{dt} &= \boldsymbol{A}\boldsymbol{x}(t) + \boldsymbol{B}\boldsymbol{u}(t), \qquad \boldsymbol{x}(0) = \boldsymbol{x}_0\\ \boldsymbol{y}(t) &= \boldsymbol{C}\boldsymbol{x}(t) + \boldsymbol{D}\boldsymbol{u}(t). &(1.18)\end{aligned}$$

The solution of this system of linear differential equations (1.18) is

$$\boldsymbol{y}(t) = \boldsymbol{C}e^{\boldsymbol{A}t}\boldsymbol{x}_0 + \boldsymbol{C}\int_0^t e^{\boldsymbol{A}(t-\tau)}\boldsymbol{B}\boldsymbol{u}(\tau)d\tau + \boldsymbol{D}\boldsymbol{u}(t) \tag{1.19}$$

which is essentially the same result as for a simple first-order differential equation with constant coefficients. The scalar factors are replaced by matrices and the exponential function is replaced by a matrix expression which will be explained later. A difference equation of an nth-order discrete-time system is rewritten as a first-order difference equation and algebraic identities. This is possible for non-linear systems as well as for linear systems. Generally, a discrete-time system is

$$\begin{aligned} \boldsymbol{x}(k+1) &= \boldsymbol{f}[\boldsymbol{x}(k), \boldsymbol{u}(k), k], \qquad \boldsymbol{x}(0) = \boldsymbol{x}_0 \\ \boldsymbol{y}(k) &= \boldsymbol{g}[\boldsymbol{x}(k), \boldsymbol{u}(k), k] \end{aligned} \tag{1.20}$$

which can be simplified in the linear case to

$$\begin{aligned} \boldsymbol{x}(k+1) &= \boldsymbol{A}\boldsymbol{x}(k) + \boldsymbol{B}\boldsymbol{u}(k), \qquad \boldsymbol{x}(0) = \boldsymbol{x}_0 \\ \boldsymbol{y}(k) &= \boldsymbol{C}\boldsymbol{x}(k) + \boldsymbol{D}\boldsymbol{u}(k). \end{aligned} \tag{1.21}$$

The solution of the system of difference equations (1.21) is

$$\boldsymbol{y}(k) = \boldsymbol{C}\boldsymbol{A}^k\boldsymbol{x}_0 + \boldsymbol{C}\sum_{i=1}^{k}\boldsymbol{A}^{i-1}\boldsymbol{B}\boldsymbol{u}(k-i) + \boldsymbol{D}\boldsymbol{u}(k). \tag{1.22}$$

Block diagrams

Taking into account the dimensions of the various vectors and matrices for an nth-order multivariable system having r inputs and m outputs with

- $\boldsymbol{x}(t), \boldsymbol{x}(k) \in \Re^{n\times 1}$ – state vector,
- $\boldsymbol{u}(t), \boldsymbol{u}(k) \in \Re^{r\times 1}$ – input vector,
- $\boldsymbol{y}(t), \boldsymbol{y}(k) \in \Re^{m\times 1}$ – output vector, controlled variable,
- $\boldsymbol{A} \in \Re^{n\times n}$ – system matrix,
- $\boldsymbol{B} \in \Re^{n\times r}$ – input matrix,
- $\boldsymbol{C} \in \Re^{m\times n}$ – output matrix,
- $\boldsymbol{D} \in \Re^{m\times r}$ – feed-through matrix,

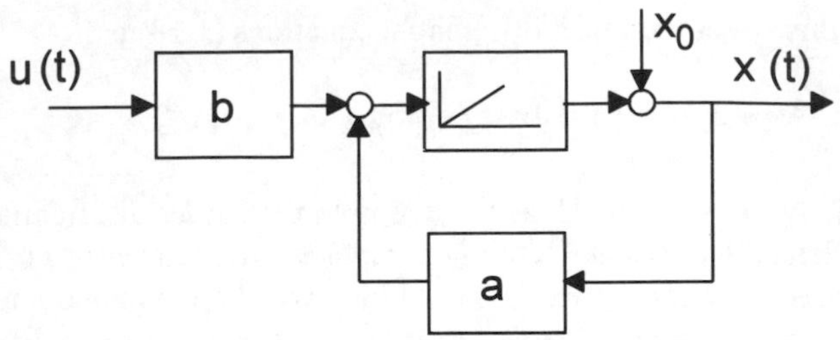

Figure 1.4. Block diagram of a first-order differential equation

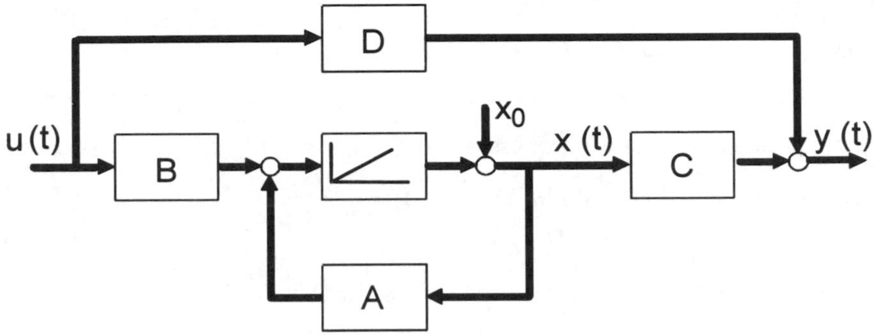

Figure 1.5. Multivariable continuous-time state-space system

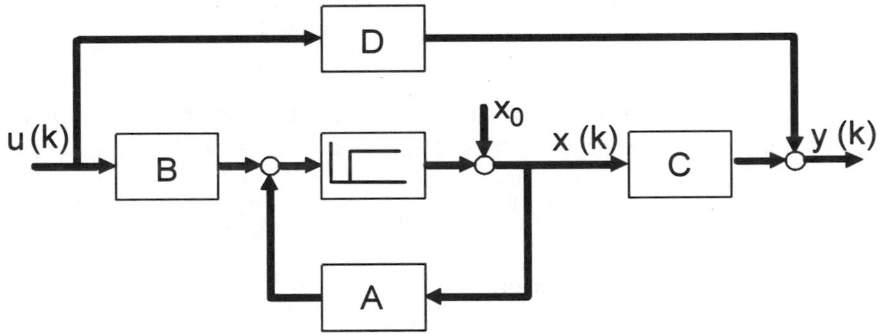

Figure 1.6. Multivariable discrete-time state-space system

one is able to construct block diagrams, which are very similar to that of a first-order differential equation of a single-input, single-output (SISO) system

$$\dot{x}(t) = ax(t) + bu(t), \qquad x(0) = x_0, \tag{1.23}$$

which is depicted in Fig. 1.4.

The block diagrams for multivariable continuous-time and discrete-time systems, which include also the output equation and direct feed-through of the input variable to the output, are illustrated in Figs 1.5 and 1.6.

In the case of a SISO system, the matrices $\boldsymbol{B}$ and $\boldsymbol{C}$ degenerate to a column and a row vector respectively, and the feed-through will become a scalar factor d, which is shown in

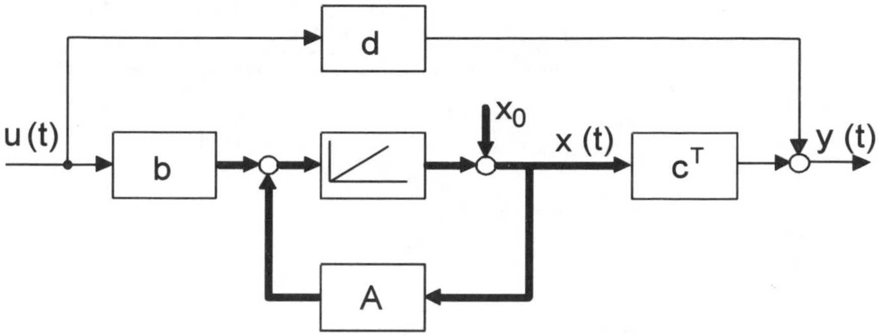

Figure 1.7. Single-input single-output state-space system

Fig. 1.7.

State-space realizations and standard forms

The structure of the matrices $\boldsymbol{A}, \boldsymbol{B}, \boldsymbol{C}, \boldsymbol{D}$ can vary widely. Canonical forms are of special meaning in the handling of state-space systems. Certain structures may be particularly useful in the numerical treatment of dynamic systems. Examples of canonical forms and/or special forms are

- controllability/controller form
- observability/observer form
- Jordan form
- Hessenberg form
- Schur form

which can be generated using similarity transformations. The Hessenberg form, which has a quasi-triangular system matrix A, arises in a stage of the transformation to the canonical forms based on a polynomial basis of the state space. These polynomial bases are structured by the Kronecker and observability/controllability indices [8] and are often used when modelling starting from an input–output description or in pole-placement design. The Jordan form and Schur form have an eigenvector basis and are very useful when dealing with eigenvalues and modal properties of a dynamical system.

Controller canonical form

As mentioned before, the controller canonical form is associated with a so-called polynomial basis of the state space. There are close relations to the characteristic polynomial and the transfer function. Consider a differential equation

$$\frac{d^n y}{dt^n} + a_{n-1}\frac{d^{n-1}y}{dt^{n-1}} + \cdots + a_1\frac{dy}{dt} + a_0 y = b_0 u + b_1\frac{du}{dt} + \cdots + b_n\frac{d^n u}{dt^n}, \tag{1.24}$$

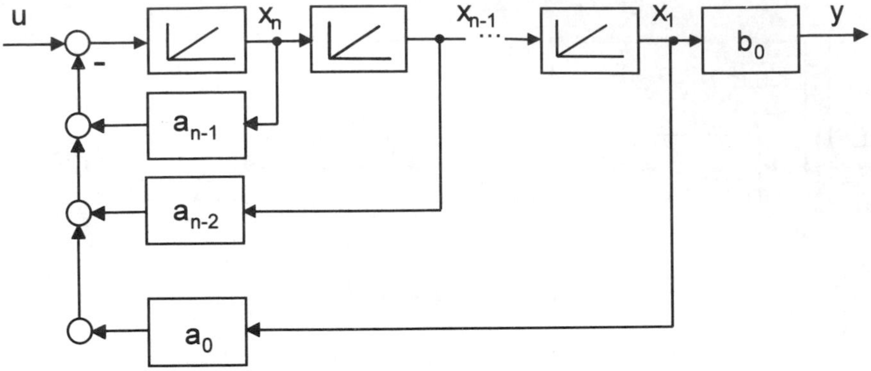

Figure 1.8. Block diagram for the first stage of realization

which has coefficients equal to those of the transfer function of the system. There are several possibilities in assigning derivatives of y to the different states variables of the state-space representation. If we first consider the special case where no derivatives of the input signal are present, as in

$$\frac{d^n y}{dt^n} + a_{n-1}\frac{d^{n-1}y}{dt^{n-1}} + \cdots + a_1\frac{dy}{dt} + a_0 y = b_0 u, \tag{1.25}$$

we can solve for the highest derivative according to

$$\frac{d^n y}{dt^n} = b_0 u - [a_{n-1}\frac{d^{n-1}y}{dt^{n-1}} + \cdots + a_1\frac{dy}{dt} + a_0 y], \tag{1.26}$$

which can be immediately interpreted in the form of a block diagram as depicted in Fig. 1.8, if the gain b_0 is shifted to the output.

If the outputs of the integrators are defined as state variables, we can write down the state equations

$$\begin{aligned}
\dot{x}_1 &= x_2 \\
\dot{x}_2 &= x_3 \\
&\vdots \\
\dot{x}_n &= u - a_0 x_1 - a_1 x_2 - \cdots - a_{n-1} x_n \\
y &= b_0 x_1 .
\end{aligned} \tag{1.27}$$

The vectors and matrices of the SISO state-space representation are

$$\begin{aligned}
\boldsymbol{A} &= \begin{bmatrix} 0 & 1 & 0 & 0 & \cdots & 0 \\ 0 & 0 & 1 & 0 & \cdots & 0 \\ 0 & 0 & 0 & 1 & \cdots & 0 \\ \vdots & \vdots & & & \ddots & \vdots \\ 0 & 0 & 0 & 0 & \cdots & 1 \\ -a_0 & -a_1 & -a_2 & -a_3 & \cdots & -a_{n-1} \end{bmatrix}, \quad \boldsymbol{b} = \begin{bmatrix} 0 \\ 0 \\ 0 \\ \vdots \\ 0 \\ 1 \end{bmatrix}, \\
\boldsymbol{c}^T &= \begin{bmatrix} b_0 & 0 & 0 & 0 & \ldots & 0 \end{bmatrix}.
\end{aligned} \tag{1.28}$$

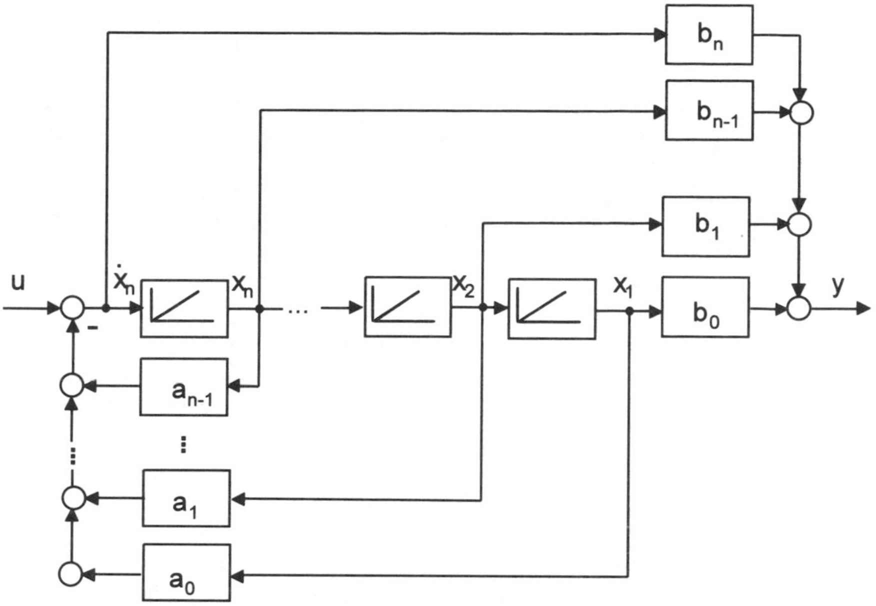

Figure 1.9. Block diagram of a SISO state-space system in controller form

The structure of the system matrix is the companion form, which means that the significant last row contains minus the coefficients of the characteristic polynomial

$$P(s) = |s\boldsymbol{I} - \boldsymbol{A}| = a_0 + a_1 s + \cdots + s^n \tag{1.29}$$

Generally, when derivatives of the input variable u are also present, this direct approach is not possible. However, if the first state variable x_1 is chosen in such a way that the output value is defined by

$$y = b_0 x_1 + b_1 \frac{dx_1}{dt} + \cdots + b_n \frac{d^n x_1}{dt^n}, \tag{1.30}$$

the same structure of the system matrix as before can be retained.

By inspection of Fig. 1.9 we can write down the state-space representation with

$$\boldsymbol{A} = \begin{bmatrix} 0 & 1 & 0 & 0 & \cdots & 0 \\ 0 & 0 & 1 & 0 & \cdots & 0 \\ 0 & 0 & 0 & 1 & \cdots & 0 \\ \vdots & \vdots & & & \ddots & \vdots \\ 0 & 0 & 0 & 0 & \cdots & 1 \\ -a_0 & -a_1 & -a_2 & -a_3 & \cdots & -a_{n-1} \end{bmatrix}, \quad \boldsymbol{b} = \begin{bmatrix} 0 \\ 0 \\ 0 \\ \vdots \\ 0 \\ 1 \end{bmatrix}, \tag{1.31}$$

$$\boldsymbol{c}^T = \begin{bmatrix} (b_0 - b_n a_0) & (b_1 - b_n a_1) & \cdots & (b_{n-1} - b_n a_{n-1}) \end{bmatrix}, \quad \boldsymbol{d} = \begin{bmatrix} b_n \end{bmatrix}. \tag{1.32}$$

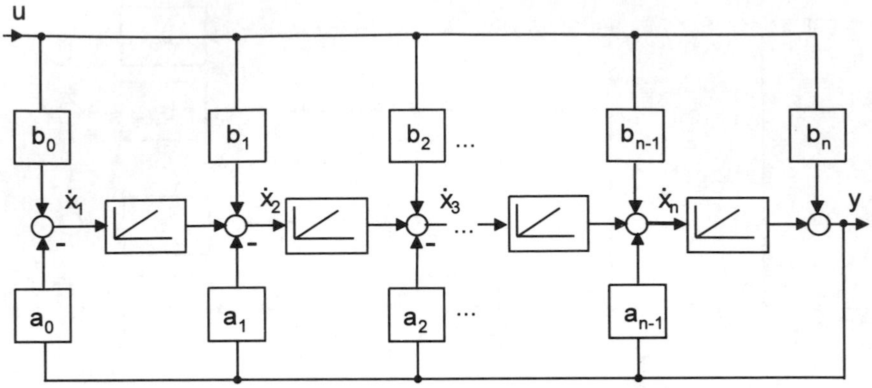

Figure 1.10. Block diagram of a SISO state-space system in observer form

Observer canonical form

The observer canonical form is dual to the controller form of state-space representation, but is introduced by a different approach. To avoid derivatives of the input value, the differential equation

$$\frac{d^n y}{dt^n} + a_{n-1}\frac{d^{n-1}y}{dt^{n-1}} + \cdots + a_1\frac{dy}{dt} + a_0 y = b_0 u + b_1\frac{du}{dt} + \cdots + b_n\frac{d^n u}{dt^n} \tag{1.33}$$

will be integrated n times, which yields

$$\begin{aligned} y(t) &= b_n u(t) + \int_0^t [b_{n-1}u(\tau) - a_{n-1}y(\tau)]d\tau + \cdots \\ &\quad + \int_0^t \cdots \int_0^t [b_0 u(\tau) - a_0 y(\tau)]d\tau^n. \end{aligned} \tag{1.34}$$

This can be realized in a block diagram as in Fig. 1.10.

If the outputs of the integrators are again defined as state variables, the state equations can be written as

$$\begin{aligned} \dot{x}_1 &= -a_0 y + b_0 u \\ \dot{x}_2 &= x_1 - a_1 + b_1 u \\ \dot{x}_3 &= x_2 - a_2 + b_2 u \\ &\vdots \\ \dot{x}_n &= x_{n-1} - a_{n-1} + b_{n-1} u \end{aligned} \tag{1.35}$$

and

$$y = x_n + b_n u. \tag{1.36}$$

Thus y can be eliminated and the state-space matrices of the observer form are

$$\boldsymbol{A} = \begin{bmatrix} 0 & 0 & \cdots & 0 & 0 & -a_0 \\ 1 & 0 & \cdots & 0 & 0 & -a_1 \\ 0 & 0 & \ddots & \vdots & \vdots & \vdots \\ \vdots & \vdots & 0 & 1 & 0 & -a_{n-2} \\ 0 & 0 & \cdots & 0 & 1 & -a_{n-1} \end{bmatrix}, \quad \boldsymbol{b} = \begin{bmatrix} b_0 - b_n a_0 \\ \vdots \\ \vdots \\ b_{n-3} - b_n a_{n-3} \\ b_{n-2} - b_n a_{n-2} \\ b_{n-1} - b_n a_{n-1} \end{bmatrix},$$

$$\boldsymbol{c}^T = \begin{bmatrix} 0 & 0 & \cdots & 0 & 1 \end{bmatrix}, \quad \boldsymbol{d} = \begin{bmatrix} b_n \end{bmatrix}. \tag{1.37}$$

The concept of observer and controller canonical form can also be applied to multivariable systems. If left-coprime polynomial fractions are used to describe the transfer-function matrix of the multivariable system, the associated multivariable canonical forms can be defined in a similar way to the approach presented here [8].

Diagonal or Jordan form

If we write the transfer function of a system having real poles with multiplicity one in terms of partial fractions

$$G(s) = \sum_{k=1}^{n} \frac{c_k}{s - \alpha_k}, \tag{1.38}$$

the output value of the system is

$$Y(s) = \sum_{k=1}^{n} \frac{c_k}{s - \alpha_k} U(s), \tag{1.39}$$

which can be represented by the block diagram in Fig. 1.11. With this definition of state variables, the state equations can be written as

$$\dot{x}_k = \alpha_k x_k + u_k, \quad k = 1, 2, ..., n. \tag{1.40}$$

The output variable is defined by

$$y = c_1 x_1 + c_2 x_2 + \cdots + c_n x_n, \tag{1.41}$$

so the matrices of the state-space representation are

$$\boldsymbol{A} = \begin{bmatrix} \alpha_1 & 0 & \cdots & 0 \\ 0 & \alpha_2 & & \vdots \\ \vdots & & \ddots & 0 \\ 0 & \cdots & 0 & \alpha_n \end{bmatrix}, \quad \boldsymbol{b} = \begin{bmatrix} 1 \\ \vdots \\ 1 \\ 1 \end{bmatrix} \tag{1.42}$$

and

$$\boldsymbol{c}^T = \begin{bmatrix} c_1 & c_2 & \cdots & c_n \end{bmatrix} \tag{1.43}$$

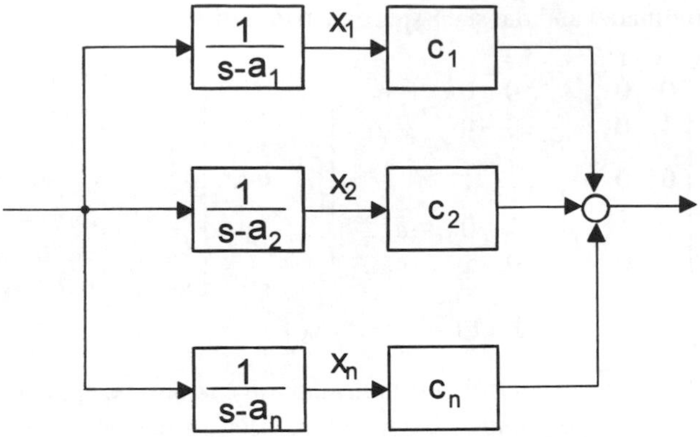

Figure 1.11. Block diagram of a state-space system in diagonal form

This quasi-modal representation may be divided into n decoupled subsystems, where each subsystem is defined by a single pole. The system matrix has diagonal form with the poles as diagonal entries.

A generalized diagonal form can be defined also for systems having a multiplicity of identical eigenvalues and complex system poles. In this case the diagonal entries will be replaced by so-called Jordan blocks, which are bidiagonal matrices with unity super- or subdiagonal entries.

1.2.3 State transformations

The last section demonstrated that there exist many possible definitions of the state variables for a given system. Thus, different matrices $\boldsymbol{A}$, $\boldsymbol{B}$ and $\boldsymbol{C}$ are obtained. On the other hand, several special forms of the state equations are often useful for special purposes. The controller canonical form is preferred for pole-placement design of state-feedback controllers, and the Jordan form can be especially useful when analysing the modes or eigenmovements of a system. Thus we need to have tools to generate these special system representations.

The definition of the states introduces a coordinate system for the n-dimensional state space. A change of this definition leads to a transformation of coordinates. So each component of the new transformed state vector can be expressed as a linear combination of the components of the original state vector. This can be written as

$$\boldsymbol{x}' = \boldsymbol{T}^{-1}\boldsymbol{x}, \quad \boldsymbol{x} = \boldsymbol{T}\boldsymbol{x}'. \tag{1.44}$$

Applying the substitution of the state vector to the state equation yields

$$\begin{aligned} \dot{\boldsymbol{x}}' &= \boldsymbol{T}^{-1}\boldsymbol{A}\boldsymbol{T}\boldsymbol{x}' + \boldsymbol{T}^{-1}\boldsymbol{B}\boldsymbol{u} = \boldsymbol{A}'\boldsymbol{x}' + \boldsymbol{B}'\boldsymbol{u} \\ \boldsymbol{y}(t) &= \boldsymbol{C}\boldsymbol{T}\boldsymbol{x}' + \boldsymbol{D}\boldsymbol{u} = \boldsymbol{C}'\boldsymbol{x}' + \boldsymbol{D}'\boldsymbol{u} \end{aligned} \tag{1.45}$$

Similarity transformations will be shortly discussed for the transformation to diagonal form. Therefore, the homogeneous system

$$\dot{\boldsymbol{x}}(t) = \boldsymbol{A}\boldsymbol{x}(t), \quad \boldsymbol{x}(0) = \boldsymbol{x}_0 \tag{1.46}$$

with matrix $\boldsymbol{A}$ having n different real eigenvalues s_i will be considered. We are looking for a transformation

$$\boldsymbol{x} = \boldsymbol{V}\boldsymbol{x}^* \tag{1.47}$$

with a full-rank transformation matrix $\boldsymbol{V}$, so that the transformed system is

$$\dot{\boldsymbol{x}}^*(t) = \boldsymbol{\Lambda}\boldsymbol{x}^*(t) \tag{1.48}$$

with

$$\boldsymbol{\Lambda} = diag[\lambda_i]. \tag{1.49}$$

The matrices $\boldsymbol{A}$ and $\boldsymbol{\Lambda}$ are similar. Both own the same eigenvalues λ_i, and

$$\boldsymbol{\Lambda} = \boldsymbol{V}^{-1}\boldsymbol{A}\boldsymbol{V}. \tag{1.50}$$

For evaluation of the transformation matrix one writes

$$\boldsymbol{A}\boldsymbol{V} = \boldsymbol{V}\boldsymbol{\Lambda} \tag{1.51}$$

and expresses this matrix equation in terms of column vectors

$$\boldsymbol{A}\begin{bmatrix} v_1 & v_2 & \cdots & v_n \end{bmatrix} = \begin{bmatrix} v_1 & v_2 & \cdots & v_n \end{bmatrix} diag[\lambda_i]. \tag{1.52}$$

Because of the diagonal form of the matrix $\boldsymbol{\Lambda}$ we obtain n independent systems of linear homogeneous equations

$$(\lambda_i \boldsymbol{I} - \boldsymbol{A})\boldsymbol{v}_i = 0, \qquad i = 1, 2, ..., n. \tag{1.53}$$

The vectors $\boldsymbol{v}_i$ are called eigenvectors of the matrix $\boldsymbol{A}$, and in the case of simple eigenvalues all of them are independent of each other, so the transformation matrix $\boldsymbol{V}$ will be non-singular.

In practice, transformations will often be done by a piecewise (or elementary) construction of the standard form. Especially for transformation to the canonical forms with a polynomial basis, elementary and orthogonal similarity operations [20] are applied.

1.2.4 Transfer function

Continuous-time signals can in many cases be represented by an nth-order differential equation of the type

$$\sum_{i=0}^{n} p_i \frac{d^i x(t)}{dt^i} = 0, \quad x^{(i)}(0) = x_0^{(i)}, \quad i = 0, 1, ..., n-1. \tag{1.54}$$

or Laplace-transformed using

$$F(s) = \int_0^\infty f(t)e^{-st}\,dt \tag{1.55}$$

as

$$\begin{aligned} X(s)\sum_{i=0}^{n} p_i s^i - \sum_{i=0}^{n-1}[x^{(i)}(0)\sum_{j=i+1}^{n} p_j s^{j-i-1}] &= 0 \\ X(s)P(s) - P^*(s) &= 0. \end{aligned} \quad (1.56)$$

Discrete-time signals can be described by nth-order linear difference equations such as

$$\sum_{i=0}^{n} p_i x(k-i) = 0, \quad i = 0, 1, ..., n-1, \quad (1.57)$$

or Laplace-transformed using

$$F^*(s) = \int_0^\infty f^*(t)e^{-st}dt, \quad (1.58)$$

with the sampled signal (with uniform sampling rate) represented by

$$\begin{aligned} f^*(t) &= f(t)\sum_{k=0}^{\infty}\delta(t-kT) = \sum_{k=0}^{\infty} f(kT)\delta(t-kT) \\ L\{f^*(t)\} &= \sum_{k=0}^{\infty} f(kt)e^{-kTs} = \sum_{k=0}^{\infty} f(kt)z^{-k} = F(z). \end{aligned} \quad (1.59)$$

Substituting

$$z = e^{Ts}, \quad F(z) = \sum_{k=0}^{\infty} z^{-k} \quad (1.60)$$

yields

$$X(z)\sum_{i=0}^{n} p_i z^{-i} - \sum_{i=0}^{n-1} x(i) \sum_{j=i+1}^{n} p_j z^{-j+i-1} = 0, \quad (1.61)$$

or equivalently

$$X(z)P(z) - P^*(z) = 0. \quad (1.62)$$

For continuous-time description by an inhomogeneous differential equation, where a dead-time T_t will be considered, one can write

$$\sum_{i=0}^{n} a_i y^{(i)}(t) = \sum_{i=0}^{m} b_i u^{(i)}(t-T_t), \quad y^{(i)}(0), i = 0, 1, ..., n-1. \quad (1.63)$$

The Laplace transform yields

$$Y(s)\sum_{i=0}^{n} a_i s^i = U(s)\sum_{i=0}^{m} b_i s^i e^{-T_t s} \quad (1.64)$$

and the transfer function (with normalized leading coefficient $a_n = 1$) is

$$G(s) = \frac{Y(s)}{U(s)} = \frac{\sum_{i=0}^{m} b_i s^i}{s^n + \sum_{i=0}^{n-1} a_i s^i} e^{-T_t s} = \frac{B(s)}{A(s)} e^{-T_t s}. \tag{1.65}$$

For a discrete-time description by an inhomogeneous difference equation

$$\sum_{i=0}^{n} a_i y(k-i) = \sum_{i=0}^{m} b_i u(k-i-d) \tag{1.66}$$

where d is an integer greater than T_t/T, the z-transform yields

$$Y(z) \sum_{i=0}^{n} a_i z^{-i} = U(z) \sum_{i=0}^{m} b_i z^{-i} z^{-d}. \tag{1.67}$$

The transfer function (with monic denominator polynomial, $a_0 = 1$) is

$$G(z) = \frac{Y(z)}{U(z)} = \frac{\sum_{i=0}^{m} b_i z^i}{1 + \sum_{i=1}^{n} a_i z^i} z^{-d} = \frac{B(z)}{A(z)} z^{-d}. \tag{1.68}$$

Slightly different representations using poles and zeros are possible, as in

$$G(p) = \frac{Y(p)}{U(p)} = V \frac{\Pi_{i=1}^{m} (p - p_{zi})}{1 + \Pi_{i=1}^{n} (p - p_{pi})} \gamma \tag{1.69}$$

with $p = s$, $\gamma = e^{-Ts}$ for continuous-time systems and $p = z$, $\gamma = z^{-d}$ for discrete-time systems. The scalar case can easily be generalized using the concept of the transfer-function matrix to the multivariable case.

1.3 SOLUTION OF THE STATE EQUATION

1.3.1 The homogenous solution – transition matrix

During the discussion of the state-space representation in the last section, its solution was defined by simply pointing out the equivalence to a first-order differential equation with constant coefficients. This will now be investigated more deeply. The first-order forced differential equation

$$\dot{x}(t) = ax(t) + bu(t), \tag{1.70}$$

with initial condition $x(0) = x_0$ can be easily solved by application of Laplace transformation, which yields

$$sX(s) - x_0 = aX(s) + bU(s). \tag{1.71}$$

By simple algebra,

$$X(s) = \frac{1}{s-a} x_0 + \frac{1}{s-a} bU(s). \tag{1.72}$$

By back-transformation to the time domain, the final solution is

$$x(t) = e^{at}x_0 + \int_0^t e^{a(t-\tau)}bu(\tau)d\tau. \tag{1.73}$$

In the case of a vector differential equation, where the constant coefficients are replaced by matrices, the formal solution can be equivalently written in terms of free and forced response as

$$\boldsymbol{x}(t) = e^{\boldsymbol{A}t}\boldsymbol{x}_0 + \int_0^t e^{\boldsymbol{A}(t-\tau)}\boldsymbol{B}\boldsymbol{u}(\tau)d\tau, \tag{1.74}$$

where the only undefined term is the matrix exponential function $e^{\boldsymbol{A}t}$, which must, by analogy to the scalar case, satisfy the identity

$$\frac{d}{dt}e^{\boldsymbol{A}t} = \boldsymbol{A}e^{\boldsymbol{A}t}. \tag{1.75}$$

A matrix with this property can be produced by carrying the series expansion of the scalar exponential function to the matrix case. Then $e^{\boldsymbol{A}t}$ can be defined as the infinite series

$$e^{\boldsymbol{A}t} = \boldsymbol{I} + \boldsymbol{A}t + \boldsymbol{A}^2\frac{t^2}{2!} + \boldsymbol{A}^3\frac{t^3}{3!} + \cdots = \sum_{k=0}^{\infty}\boldsymbol{A}^k\frac{t^k}{k!}, \tag{1.76}$$

which converges for all matrices $\boldsymbol{A}$ and $|t| < \infty$ absolutely. Differentiating the series with respect to time,

$$\begin{aligned}\frac{d}{dt}e^{\boldsymbol{A}t} &= \boldsymbol{A} + \boldsymbol{A}^2t + \boldsymbol{A}^3\frac{t^2}{2!} + \cdots \\ &= \boldsymbol{A}\left[\boldsymbol{I} + \boldsymbol{A}t + \boldsymbol{A}^2\frac{t^2}{2!} + +\boldsymbol{A}^3\frac{t^3}{3!} + \cdots\right] = \boldsymbol{A}e^{\boldsymbol{A}t}\end{aligned} \tag{1.77}$$

By substituting the solution into the original state equation and differentiating with respect to time t we obtain

$$\begin{aligned}\dot{\boldsymbol{x}}(t) &= \boldsymbol{A}e^{\boldsymbol{A}t}\boldsymbol{x}_0 + \boldsymbol{A}e^{At}\int_0^t e^{\boldsymbol{A}t}\boldsymbol{B}\boldsymbol{u}(t-\tau)d\tau + e^{\boldsymbol{A}t}e^{-\boldsymbol{A}t}\boldsymbol{B}\boldsymbol{u}(t) \\ &= \boldsymbol{A}\left[e^{\boldsymbol{A}t}\boldsymbol{x}_0 + \int_0^t e^{\boldsymbol{A}(t-\tau)}\boldsymbol{B}\boldsymbol{u}(t-\tau)d\tau\right] + \boldsymbol{B}\boldsymbol{u}(t) \\ &= \boldsymbol{A}\boldsymbol{x}(t) + \boldsymbol{B}\boldsymbol{u}(t).\end{aligned} \tag{1.78}$$

The matrix $e^{\boldsymbol{A}t}$ is often called the transition matrix and the special symbol $\boldsymbol{\Phi}(t)$ is used for this matrix function.

1.3.2 Calculation of the transition matrix

Eigenvalue method

For a diagonal system matrix the transition matrix is

$$\boldsymbol{\Phi}^* = e^{\boldsymbol{A}t} = diag[e^{s_i}]. \tag{1.79}$$

which can be constructed using a similarity transformation. Thus the transition matrix for a general state-space system can be calculated by transforming the system to diagonal or Jordan form using

$$\boldsymbol{\Lambda} = \boldsymbol{V}^{-1}\boldsymbol{A}\boldsymbol{V}, \tag{1.80}$$

writing down the solution of the diagonalized equations in the transformed coordinate system and transforming back according to

$$\boldsymbol{\Phi}(t) = e^{\boldsymbol{A}t} = \boldsymbol{V}\boldsymbol{\Phi}^*(t)\boldsymbol{V}^{-1} = \boldsymbol{V}e^{\boldsymbol{\Lambda}t}\boldsymbol{V}^{-1}. \tag{1.81}$$

Cayley–Hamilton method

The Cayley–Hamilton theorem [8] says that each square matrix $\boldsymbol{A}$ satisfies its characteristic polynomial.

If the characteristic polynomial

$$P(s) = |s\boldsymbol{I} - \boldsymbol{A}| = a_0 + a_1 s + a_2 s^2 + \cdots + s^n = 0 \tag{1.82}$$

is known, we can write, according to the Cayley-Hamilton theorem,

$$\boldsymbol{P}(\boldsymbol{A}) = a_0\boldsymbol{I} + a_1\boldsymbol{A} + a_2\boldsymbol{A}^2 + \cdots + \boldsymbol{A}^n = \boldsymbol{0}, \tag{1.83}$$

where $\boldsymbol{P}(\boldsymbol{A})$ is an $n \times n$ matrix. Defining

$$\boldsymbol{F}(\boldsymbol{A}) = f_0\boldsymbol{I} + f_1\boldsymbol{A} + f_2\boldsymbol{A}^2 + \cdots + \boldsymbol{A}^p, \tag{1.84}$$

with $p > n$, we can write

$$\boldsymbol{F}(\boldsymbol{A}) = \boldsymbol{Q}(\boldsymbol{A})\boldsymbol{P}(\boldsymbol{A}) + \boldsymbol{R}(\boldsymbol{A}), \tag{1.85}$$

which can be interpreted in terms of polynomial division. Keeping in mind that $\boldsymbol{P}(\boldsymbol{A}) = \boldsymbol{0}$, one obtains

$$\boldsymbol{F}(\boldsymbol{A}) = \boldsymbol{R}(\boldsymbol{A}) = \alpha_0\boldsymbol{I} + \alpha_1\boldsymbol{A} + \alpha_2\boldsymbol{A}^2 + \cdots + \alpha_{n-1}\boldsymbol{A}^{n-1}. \tag{1.86}$$

Thus, as consequence of the Cayley–Hamilton theorem, we can state that

- Each $n \times n$ matrix function $\boldsymbol{F}(\boldsymbol{A})$ of degree $p \geq n$ can be expressed as a function of maximum degree $n - 1$.

This is also valid for an infinite series, if the limit as $p \to \infty$ exists. Thus we have a method for the evaluation of the matrix exponential $\boldsymbol{F}(\boldsymbol{A}) = e^{\boldsymbol{A}t}$, and the transition matrix can be expressed in the form

$$\boldsymbol{\Phi}(t) = e^{\boldsymbol{A}t} = \alpha_0(t)\boldsymbol{I} + \alpha_1(t)\boldsymbol{A} + \alpha_2(t)\boldsymbol{A}^2 + \cdots + \alpha_{n-1}(t)\boldsymbol{A}^{n-1} = \boldsymbol{R}(\boldsymbol{A}) \tag{1.87}$$

where the coefficients a_j are time functions. If we substitute the eigenvalues s_i of the matrix $\boldsymbol{A}$, we obtain n equations of the form

$$e^{s_i t} = \alpha_0(t) + \alpha_1(t)s_i + \alpha_2(t)s_i^2 + \cdots + \alpha_{n-1}(t)s_i^{n-1}, \tag{1.88}$$

if we suppose that all n eigenvalues are different from each other.

Laplace transform method

If the state equations are Laplace transformed, we obtain

$$\begin{aligned} s\boldsymbol{X}(s) - \boldsymbol{x}(0) &= \boldsymbol{A}\boldsymbol{X}(s) + \boldsymbol{B}\boldsymbol{U}(s) \\ \boldsymbol{Y}(s) &= \boldsymbol{C}\boldsymbol{X}(s) + \boldsymbol{B}\boldsymbol{U}(s), \end{aligned} \tag{1.89}$$

where the state equation can be rewritten as

$$(s\boldsymbol{I} - \boldsymbol{A})\boldsymbol{X}(s) = \boldsymbol{x}(0) + \boldsymbol{B}\boldsymbol{U}(s), \tag{1.90}$$

or equivalently as

$$\boldsymbol{X}(s) = (s\boldsymbol{I} - \boldsymbol{A})^{-1}\boldsymbol{x}(0) + (s\boldsymbol{I} - \boldsymbol{A})^{-1}\boldsymbol{B}\boldsymbol{U}(s), \tag{1.91}$$

which is allowed because $(s\boldsymbol{I} - \boldsymbol{A})$ is non-singular. By comparing this solution and the time-domain solution, we can conclude that the transition matrix is

$$\boldsymbol{\Phi}(t) = \mathcal{L}^{-1}\{(s\boldsymbol{I} - \boldsymbol{A})^{-1}\} = \mathcal{L}^{-1}\left\{\frac{1}{|s\boldsymbol{I} - \boldsymbol{A}|}\mathrm{adj}(s\boldsymbol{I} - \boldsymbol{A})\right\}. \tag{1.92}$$

1.4 SYSTEM STABILITY

1.4.1 Stability concepts and definitions

Input–output stability

The theory of stability of dynamical systems arose from problems in theoretical mechanics and has since been applied to problems of control engineering. For the type of linear time-invariant systems solely discussed here, a large number of methods to test for stability are known and these are often rather simple to handle. The basis of these methods is the definition of stability or input–output stability, which will be defined now.

A linear time-invariant system with input variable $u(t)$ and output variable $y(t)$ will be considered, and as usual these time functions will only be examined over the time interval $t \geq 0$. Such a transfer system will be said to be input–output stable if it responds to every bounded input variable $u(t)$ with a bounded output variable. This is normally termed BIBO (bounded input, bounded output) stability. Direct use of this stability definition to test stability of linear time-invariant systems is often not possible, because the class of permissible input time functions $u(t)$ is too large. As a result, often only pulse or step functions are considered. A linear system which is stable for step inputs will be generally BIBO-stable.

Zero-state stability

The idea of zero-state stability arises in connection with the controllability and observability of linear systems. In some cases an input–output representation of the system does not fully describe the system; there may exist hidden modes, i.e. states which are either uncontrollable or unobservable or both. Consider a second-order electrical system which consists of two capacitors connected in parallel. It will not be possible to apply a different charge to each of them by any input voltage. This is not a completely controllable system and its input–output behaviour is that of a first-order system. Instead of this marginally stable example, we could also think of a system having unstable hidden modes, which is defined as non-stabilizable. Based on the input–output description, we cannot decide whether this system is stable, because the unstable eigenvalue would be cancelled by an unstable zero. But the state-space description of the system contains information about unstable hidden modes and it would be better to use a different definition for stability from BIBO-stability for state-space systems.

Zero-state stability means, for a linear state-space system (the unforced case with initial condition $\boldsymbol{x}_0$), that the trajectory described by the state vector will reach the origin of the state space $\boldsymbol{x} = 0$ in a finite time.

1.4.2 Routh–Hurwitz criterion

A linear continuous-time system will be stable if all its poles have negative real parts. To test the stability of a linear systems with transfer function

$$G(s) = \frac{B(s)}{A(s)} = \frac{b_0 + b_1 s + \cdots + b_m s^m}{a_0 + a_1 s + \cdots + a_n s^n} \tag{1.93}$$

it is sufficient to check the position of the poles of the transfer function $G(s)$, i.e. the zeros s_i of the characteristic equation

$$P(s) = a_0 + a_1 s + \cdots + a_n s^n = 0. \tag{1.94}$$

The system is asymptotically stable if all poles s_i are situated in the left half plane. It is marginally stable if no pole is in the right plane, no multiple poles are on the imaginary axis and at least a single pole on the imaginary axis exists. Using stability criteria, [4], we can easily evaluate whether the characteristic polynomial $P(s)$ refers to a stable system. Necessary and sufficient conditions can be derived from a special matrix, the n-row Hurwitz matrix

$$\boldsymbol{H}_n = \begin{bmatrix} a_{n-1} & a_{n-3} & a_{n-5} & \cdots & 0 \\ a_n & a_{n-2} & a_{n-4} & \cdots & 0 \\ 0 & a_{n-1} & a_{n-3} & \cdots & 0 \\ 0 & a_n & a_{n-2} & \cdots & 0 \\ 0 & 0 & a_{n-1} & \cdots & 0 \\ \vdots & \vdots & \vdots & & \vdots \\ 0 & 0 & 0 & \cdots & a_0 \end{bmatrix}. \tag{1.95}$$

Its entries h_{ij}, the coefficients a_i of the characteristic polynomial $P(s)$, can be calculated from

$$\begin{aligned} h_{ij} &= a_{n+i-2j}, \qquad 0 \leq 2j-1 \leq n \\ h_{ij} &= 0, \qquad i > j, \quad 2j-i > n. \end{aligned} \tag{1.96}$$

To check stability, all principal minors, i.e. all north-western subdeterminants of $\boldsymbol{H}_n$, have to be generated. These are the determinants $d_1, d_2, \ldots, d_n$ as in Fig. 1.12. The original formulation of the Hurwitz criterion is as follows:

The necessary and sufficient condition for the roots of $P(s)$ to have negative real parts,

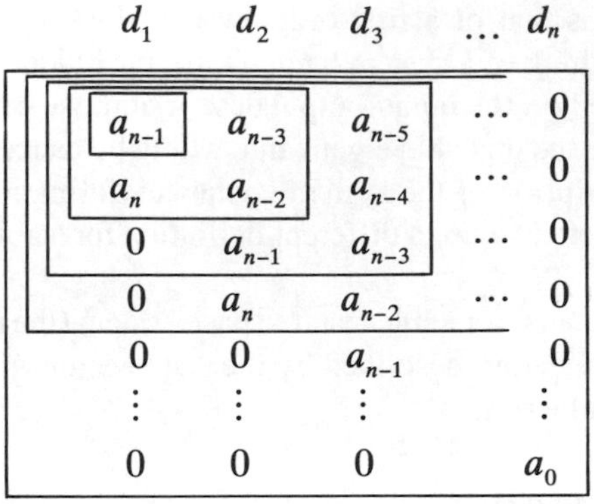

Figure 1.12. Formation of Hurwitz determinants

where the coefficient a_0 is positive, is that the determinants $d_i, i = 1, 2, ..., n$ are all positive.

This can be evaluated simply by forming the Hurwitz matrix and calculating and testing the subdeterminants, which can be done using any linear algebra subroutine package by applying LU factorization or the numerically more reliable singular-value decomposition. The Hurwitz formulation should be preferred over the Routh array, which could also be interpreted as evaluation of numerically non-stabilized Gaussian elimination for determining a triangular form of the Hurwitz matrix, when implementing a numerical test.

1.4.3 Lyapunov's method

It will not be possible to discuss all aspects of Lyapunov's stability theory in detail. Lyapunov's second method is especially useful for general non-linear systems, but here we are dealing only with linear systems and our aim is to check a system for global asymptotic stability. So only one statement of Lyapunov's theory will be given here and a method to test the stability of a standard state-space system will shortly be discussed.

The system

$$\dot{\boldsymbol{x}} = \boldsymbol{f}(\boldsymbol{x}) \tag{1.97}$$

is supposed to have an equilibrium point $\boldsymbol{x} = \mathbf{0}$. The equilibrium is globally asymptotically stable if a function $V(\boldsymbol{x})$ exists with the following properties:

1. $V(\boldsymbol{x})$ and $\boldsymbol{\nabla} V(\boldsymbol{x})$ are continuous,

2. $V(\boldsymbol{x})$ is positive definite, i.e. strictly positive except at $\boldsymbol{x} = \mathbf{0}$,

3. $\dot{V}(\boldsymbol{x}) = [\boldsymbol{\nabla} V(\boldsymbol{x})]^T \boldsymbol{f}(\boldsymbol{x})$ is negative definite,

4. $lim_{\|\boldsymbol{x}\| \to \infty} V(\boldsymbol{x}) = \infty$.

Generally it will be difficult to find a suitable Lyapunov function $V(\boldsymbol{x})$. For a linear system

$$\dot{\boldsymbol{x}} = \boldsymbol{A}\boldsymbol{x}(t) \tag{1.98}$$

a Lyapunov function can be obtained as the quadratic form

$$V(\boldsymbol{x}) = \boldsymbol{x}^T \boldsymbol{P} \boldsymbol{x}, \tag{1.99}$$

with a positive-definite matrix $\boldsymbol{P}$. The time derivative is

$$\dot{V}(\boldsymbol{x}) = \dot{\boldsymbol{x}}^T \boldsymbol{P} \boldsymbol{x} + \boldsymbol{x}^T \boldsymbol{P} \dot{\boldsymbol{x}}. \tag{1.100}$$

Using the transposed state equation $\dot{\boldsymbol{x}}^T = \boldsymbol{x}^T \boldsymbol{A}^T$, we obtain

$$\dot{V}(\boldsymbol{x}) = \boldsymbol{x}^T \left[\boldsymbol{A}^T \boldsymbol{P} + \boldsymbol{P}^T \boldsymbol{A}\right] \boldsymbol{x}, \tag{1.101}$$

which is again a quadratic form, which has to be negative definite for asymptotic stability. Define a positive-definite matrix $\boldsymbol{Q}$ satisfying the matrix Lyapunov equation

$$\left[\boldsymbol{A}^T \boldsymbol{P} + \boldsymbol{P}^T \boldsymbol{A}\right] = -\boldsymbol{Q}. \tag{1.102}$$

Application of the Lyapunov theorem yields the following: if the equilibrium point $\boldsymbol{x} = 0$ is asymptotically globally stable, there exists for each positive-definite matrix $\boldsymbol{Q}$ a positive-definite matrix $\boldsymbol{P}$ which satisfies the matrix Lyapunov equation. We can define a positive-definite $\boldsymbol{Q}$, solve the Lyapunov equation for $\boldsymbol{P}$ and check the stability by testing whether $\boldsymbol{P}$ is positive definite. Global asymptotic stability means that all eigenvalues of $\boldsymbol{A}$ have negative real parts.

1.5 CONTROLLABILITY AND OBSERVABILITY

Some further properties such as reachability and reconstructability are used to examine how a system behaves for certain control inputs, looking particularly at whether the state vector moves to or away from the origin of the state space. These properties will not be considered here; only controllability and observability will be introduced, which are important for the design of state-feedback controllers and state observers.

1.5.1 Controllability

Definition

Linear continuous-time and discrete-time multivariable state-space systems

$$\begin{aligned} \frac{d\boldsymbol{x}(t)}{dt} &= \boldsymbol{A}\boldsymbol{x}(t) + \boldsymbol{B}\boldsymbol{u}(t), \qquad \boldsymbol{x}(0) = \boldsymbol{x}_0 \\ \boldsymbol{y}(t) &= \boldsymbol{C}\boldsymbol{x}(t) + \boldsymbol{D}\boldsymbol{u}(t) \end{aligned} \tag{1.103}$$

$$\begin{aligned} \boldsymbol{x}(k+1) &= \boldsymbol{A}\boldsymbol{x}(k) + \boldsymbol{B}\boldsymbol{u}(k), \qquad \boldsymbol{x}(0) = \boldsymbol{x}_0 \\ \boldsymbol{y}(k) &= \boldsymbol{C}\boldsymbol{x}(k) + \boldsymbol{D}\boldsymbol{u}(k) \end{aligned} \tag{1.104}$$

will be considered.

Definitions:
The system in Eq. (1.103) is said to be completely state controllable if, for any given initial

state $\boldsymbol{x}_0$, there exists a time t_f and a control value $\boldsymbol{u}(t)$ defined in the interval $[0, t_f]$ such that $\boldsymbol{x}(t_f) = \mathbf{0}$.

The system in Eq. (1.104) is said to be completely state controllable if, for any given initial state $\boldsymbol{x}_0$, there exists an integer $p > 0$ and a sequence $\boldsymbol{u}(k)$ in the interval $[0, p-1]$ such that $\boldsymbol{x}_p = \mathbf{0}$.

Criteria

As in the definition of stability, the definition of controllability cannot be used directly to test whether a system is controllable. We need necessary and sufficient criteria which can be handled more easily mathematically and numerically.

The system in Eq. (1.103) is controllable if and only if any one of the following conditions is satisfied:

1. $\text{rank}[\boldsymbol{B}, \boldsymbol{AB}, \ldots, \boldsymbol{A}^{n-1}\boldsymbol{B}] = n$;

2. $\text{rank}[\boldsymbol{B}, \boldsymbol{AB}, \ldots, \boldsymbol{A}^{q-r}\boldsymbol{B}] = n, r = \text{rank}\boldsymbol{B}$ and q is the degree of the so-called minimal polynomial of the system matrix $\boldsymbol{A}$;

3. $\text{rank}[s\boldsymbol{I} - \boldsymbol{A}, \boldsymbol{B}] = n$ for $s \in \mathcal{C}$, where $\mathcal{C}$ denotes the field of complex numbers;

4. $\text{rank}[s\boldsymbol{I} - \boldsymbol{A}, \boldsymbol{B}] = n$ for $s \in \sigma_A$, where σ_A denotes the spectrum of $\boldsymbol{A}$;

5. the (polynomial) matrices $s\boldsymbol{I} - \boldsymbol{A}$ and $\boldsymbol{B}$ are left coprime;

6. the rows of $[s\boldsymbol{I} - \boldsymbol{A}]^{-1}\boldsymbol{B}$ are linearly independent over $\Re(s)$, the field of rational functions;

7. the rows of $e^{\boldsymbol{A}t}\boldsymbol{B}$ are linearly independent;

8. the matrix $\boldsymbol{W}(t_0, t_f) = \int_{t_0}^{t_f} e^{\boldsymbol{A}t}\boldsymbol{B}\boldsymbol{B}^T e^{\boldsymbol{A}^T t} dt, t_f > t_0$ is positive definite;

9. the matrix $\boldsymbol{W}(t_0, t_f)$ is non-singular;

10. $\boldsymbol{x} \in R[\boldsymbol{B}, \boldsymbol{AB}, \ldots, \boldsymbol{A}^{n-1}\boldsymbol{B}]$ for all $\boldsymbol{x} \in \Re^n$, where R denotes the range space, i.e. the collection of all linear combinations of the columns of the matrix $[\boldsymbol{B}, \boldsymbol{AB}, \ldots, \boldsymbol{A}^{n-1}\boldsymbol{B}]$.

All conditions above involving rank tests of real-valued matrices can be used to test controllability of discrete-time state-space systems as given in Eq. (1.104). In conditions 3 – 6 the variable s must be replaced by z. Conditions 8 and 9, which depend on functions in the time domain, have to be replaced by

8′. the matrix $\boldsymbol{W}_D = \sum_{k=0}^{n-1} \boldsymbol{A}_k \boldsymbol{B}\boldsymbol{B}^T (\boldsymbol{A}^T)^k$ is positive definite;

9′. the matrix $\boldsymbol{W}_D$ is non-singular.

The best-known condition, mentioned in nearly all textbooks on control theory, is 1, where the Kalman controllability matrix is formed and checked for linearly dependent columns. The question of which is the best criterion to use generally cannot be answered definitely. It depends on the structure and data of the problem. All rank tests using powers of matrices are numerically poor. Thus, if numerical algorithms to test controllability are to be used, conditions 1 or 2 are not well suited, but they are simple to implement on a digital computer. The utilization of polynomial matrix equivalence transformations to implement the Popov–Belevitch–Hautus rank test for polynomial matrices (4 or 5) will also lead to very poor numerical results.

Consider the system with

$$\boldsymbol{A} = \begin{bmatrix} 0 & 1 \\ 0 & 0 \end{bmatrix}, \boldsymbol{B} = \begin{bmatrix} 1 \\ 0 \end{bmatrix}. \tag{1.105}$$

Because of the special structure of the system matrix, the repeated zero eigenvalue of $\boldsymbol{A}$ can be taken directly from the diagonal of the upper-triangular matrix. Using condition 4 one can conclude without any further calculation that the system $\{\boldsymbol{A}, \boldsymbol{B}\}$ is not controllable.

As demonstrated by this trivial example, it is often better to generate the special structure of the state-space matrices, so that one of the criteria for controllability can be applied by a simple inspection of the transformed matrices. The numerically most useful criterion is the rank of $[s\boldsymbol{I} - \boldsymbol{A}, \boldsymbol{B}]$, where the system $\{\boldsymbol{A}, \boldsymbol{B}\}$ is in a special generalized Hessenberg form, which can be generated by numerically stabilized elementary or orthogonal similarity transformations.

1.5.2 Observability

Definition

The definition of observability, which means that by observation of measurable input and output signals the state can be reconstructed in a finite time, is quite similar to the definition of controllability.

Definitions:
The system in Eq. (1.103) is said to be completely state observable if there exists a time t_f such that from given $\boldsymbol{u}(t)$ and $\boldsymbol{y}(t)$ in the interval $[0, t_f]$ we can reconstruct the initial state $\boldsymbol{x}_0$.

The system in Eq. (1.104) is said to be completely state observable if there exists an integer $q > 0$, such that from given sequences $\{\boldsymbol{u}_k\}$ and $\{\boldsymbol{y}_k\}, k = 0, \ldots, q-1$ we can reconstruct the initial state $\boldsymbol{x}_0$.

Criteria

The criteria used in testing observability are dual to the controllability criteria. The continuous-time state-space system in Eq. (1.103) is observable if and only if any one of the following conditions is satisfied:

1. $\text{rank}[\boldsymbol{C}^T, \boldsymbol{C}^T\boldsymbol{A}^T, \ldots, \boldsymbol{C}^T(\boldsymbol{A}^{n-1})^T]^T = n$;

2. $\text{rank}[C^T, C^T A^T, \ldots, C^T (A^{q-r})^T]^T = n$, $r = \text{rank}[C]$ and q is the degree of the minimal polynomial of the system matrix A;

3. $\text{rank}[sI - A^T, C^T]^T = n$, for $s \in \mathcal{C}$;

4. $\text{rank}[sI - A^T, C^T]^T = n$, for $s \in \sigma_A$;

5. the (polynomial) matrices $[sI - A]$ and C are right coprime;

6. the columns of $C[sI - A]^{-1}$ are linearly independent over $\Re(s)$;

7. the columns of Ce^{At} are linearly independent;

8. the matrix $R(t_0, t_f) = \int_{t_0}^{t-f} e^{A^T t} C^T C e^{At} dt$, $t_f > t_0$ is positive definite;

9. the matrix $R(t_0, t_f)$ is non-singular;

10. $\ker[C^T, C^T A^T, \ldots, C^T (A^{n-1})^T]^T = 0$, where ker denotes the kernel space.

For discrete-time systems no different conditions need be formulated if only rank tests of real-valued matrices are used. A change of the variable s to z is necessary in the conditions formulated using polynomial or rational matrices. The observability Grammian of the discrete-time system is $R_D = \sum_{k=0}^{n-1} (A^T)^k C^T C A^k$.

As is obvious from the conditions in earlier sections, all definitions and conditions for controllability and observability are very similar to each other, i.e. observability and controllability are dual properties.

Controllability–observability duality

As for stability criteria for transfer functions, where the use of stability tests for continuous-time and discrete-time systems can be interchanged by introducing a bilinear transformation which maps the left half plane to the unit circle, it is possible in the case of the criteria for controllability/observability to make things easier by using the concept of the dual system. In this case one has only to implement algorithms to test for controllability of a system and the check for observability can be carried out with the same algorithms applied to the dual system.

If we inspect the conditions formulated in previous sections, we can conclude that the test for observability can be carried over to a test on controllability of a system with system matrix A^T and input matrix $B = C^T$. Or, in terms of Rosenbrock's system matrix [19]

$$P = \begin{bmatrix} A & B \\ C & D \end{bmatrix}, \tag{1.106}$$

the dual system is just given by the matrix

$$\tilde{P} = \begin{bmatrix} A^T & C^T \\ B^T & D^T \end{bmatrix}. \tag{1.107}$$

1.6 FEEDBACK CONTROL AND STATE ESTIMATION

The state-space description of a linear, time-invariant, dynamic, multivariable system is slightly augmented by incorporating a disturbance term, so that it is represented in continuous time by

$$\begin{aligned}\frac{d\boldsymbol{x}(t)}{dt} &= \boldsymbol{A}\boldsymbol{x}(t) + \boldsymbol{B}\boldsymbol{u}(t) + \boldsymbol{B}_z\boldsymbol{z}_s(t)\\ \boldsymbol{y}(t) &= \boldsymbol{C}\boldsymbol{x}(t) + \boldsymbol{D}\boldsymbol{u}(t)\end{aligned} \tag{1.108}$$

and in discrete time by

$$\begin{aligned}\boldsymbol{x}(k+1) &= \boldsymbol{A}\boldsymbol{x}(k) + \boldsymbol{B}\boldsymbol{u}(k) + \boldsymbol{B}_z\boldsymbol{z}_s(k)\\ \boldsymbol{y}(k) &= \boldsymbol{C}\boldsymbol{x}(k) + \boldsymbol{D}\boldsymbol{u}(k),\end{aligned} \tag{1.109}$$

with disturbance vector $\boldsymbol{z}_s(t), \boldsymbol{z}_s(k) \in \Re^{l\times 1}$ and disturbance input matrix $\boldsymbol{B}_z \in \Re^{n\times l}$.

1.6.1 Pole placement

Considering an nth-order SISO system with the state-space description

$$\dot{\boldsymbol{x}}(t) = \boldsymbol{A}\boldsymbol{x}(t) + \boldsymbol{b}u(t), \tag{1.110}$$

the problem of pole placement using state feedback

$$u = -\boldsymbol{f}^T\boldsymbol{x} \tag{1.111}$$

will be handled shortly. The aim of this approach is the calculation of the feedback vector $\boldsymbol{f}$, so that the matrix

$$\boldsymbol{A}_g = \boldsymbol{A} - \boldsymbol{b}\boldsymbol{f}^T \tag{1.112}$$

of the closed-loop system has the pre-specified poles

$$p_1, p_2, \ldots, p_n \tag{1.113}$$

as eigenvalues. Calculation of $\boldsymbol{f}$ will be much simplified if the system $\{\boldsymbol{A}, \boldsymbol{b}\}$ is in controller canonical form

$$\boldsymbol{A} = \begin{bmatrix} 0 & 1 & 0 & 0 & \cdots & 0\\ 0 & 0 & 1 & 0 & \cdots & 0\\ 0 & 0 & 0 & 1 & \cdots & 0\\ \vdots & \vdots & & & \ddots & \vdots\\ 0 & 0 & 0 & 0 & \cdots & 1\\ -a_0 & -a_1 & -a_2 & -a_3 & \cdots & -a_{n-1}\end{bmatrix}, \quad \boldsymbol{b} = \begin{bmatrix}0\\0\\0\\ \vdots\\0\\1\end{bmatrix}. \tag{1.114}$$

In this canonical representation, matrix $\boldsymbol{A}_g$ remains in controller canonical form, and the entries of $\boldsymbol{f}$ are simply

$$f_i = q_{i-1} - a_{i-1}, \tag{1.115}$$

where q_i are the coefficients of the closed-loop characteristic polynomial

$$q(p) = \Pi_{i=1}^{n}(p - p_i) = q_0 + q_1 p + q_2 p^2 + \cdots + q_{n-1}p^{n-1} + p^n . \quad (1.116)$$

If the system is not given in controller canonical form, we have to transform it using a similarity transformation

$$\boldsymbol{x}_R = \boldsymbol{T}\boldsymbol{x}, \quad (1.117)$$

so that the system

$$\dot{\boldsymbol{x}}_R = \boldsymbol{A}_R \boldsymbol{x}_R + b_R u = \boldsymbol{T}\boldsymbol{A}\boldsymbol{T}^{-1}\boldsymbol{x}_R + \boldsymbol{T}bu \quad (1.118)$$

is in controller canonical form. After calculation of the feedback gain vector $\boldsymbol{f}_R$, the feedback gain vector of the original system can be calculated using the back-transformation

$$\boldsymbol{f}^T = \boldsymbol{f}_R^T \boldsymbol{T}. \quad (1.119)$$

The main effort in pole assignment using the controller form consists in the transformation to the specific canonical form.

1.6.2 Control loop with state-feedback controller and observer

The design of numerous types of controllers in the time domain using state-space methods is today a standard task for the control engineer. Validated numerical algorithms for basic linear algebra and methods for analysis and design of systems in a state-space representation have become widely available.

One goal of state feedback is, by using a suitable feedback matrix $\boldsymbol{F}$, to specify the poles as well as the modes of the closed-loop system. Therefore the plant has to be a controllable system. Because the states of the plant are generally not all measurable, an observer has to be used to reconstruct the unmeasurable states. To achieve convergence of the estimated state vector to that of the plant, the estimation error $\boldsymbol{y} - \hat{\boldsymbol{y}}$ is fed back to the observer system using a matrix $\boldsymbol{H}$. All poles and modes of the estimation error system will be completely determined by the matrix H in the case of complete observability.

In the state-feedback/observer control loop of Fig. 1.13 we can interpret the feedback signal $\boldsymbol{f}(k)$ or $\boldsymbol{f}(t)$ as the output of a control system with two vector inputs $\boldsymbol{u}$ and $\boldsymbol{y}$. This can be described by the state-space system

$$\begin{aligned} \frac{d\hat{\boldsymbol{x}}}{dt} &= (\boldsymbol{A} - \boldsymbol{H}\boldsymbol{C})\hat{\boldsymbol{x}}(t) + \boldsymbol{B}\boldsymbol{u}(t) + \boldsymbol{H}\boldsymbol{y}(t) \\ \boldsymbol{f}(t) &= \boldsymbol{F}\hat{\boldsymbol{x}}(t), \end{aligned} \quad (1.120)$$

or

$$\begin{aligned} \hat{\boldsymbol{x}}(k+1) &= (\boldsymbol{A} - \boldsymbol{H}\boldsymbol{C})\hat{\boldsymbol{x}}(k) + \boldsymbol{B}\boldsymbol{u}(k) + \boldsymbol{H}\boldsymbol{y}(k) \\ \boldsymbol{f}(k) &= \boldsymbol{F}\hat{\boldsymbol{x}}(k). \end{aligned} \quad (1.121)$$

Applying the Laplace-transform to system (1.120) yields the correspondence

$$\boldsymbol{f}(s) = \boldsymbol{F}(s\boldsymbol{I} - \boldsymbol{A} + \boldsymbol{H}\boldsymbol{C})^{-1}\boldsymbol{B}\boldsymbol{u}(s) + \boldsymbol{F}(s\boldsymbol{I} - \boldsymbol{A} + \boldsymbol{H}\boldsymbol{C})^{-1}\boldsymbol{H}\boldsymbol{y}(s) \quad (1.122)$$

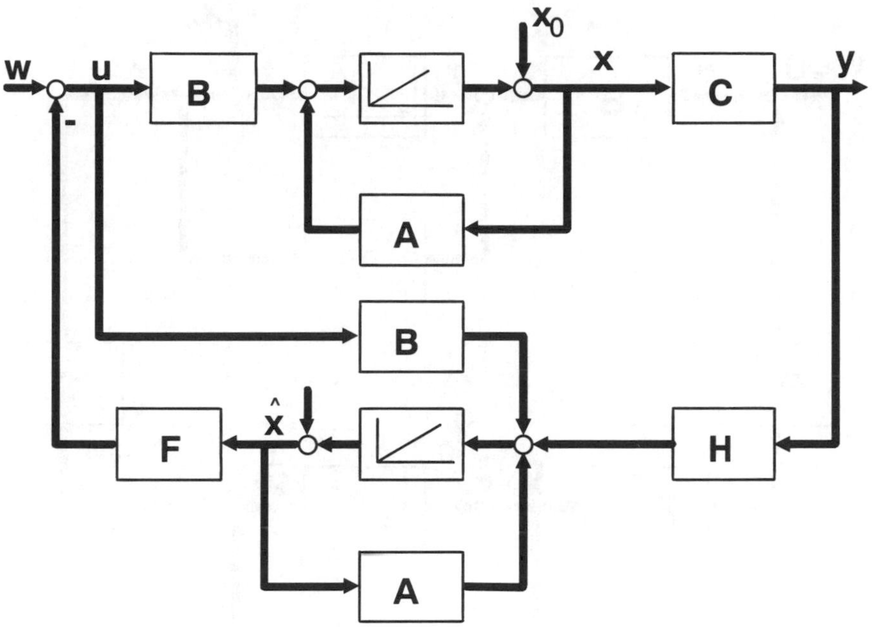

Figure 1.13. State-feedback controller/observer control loop

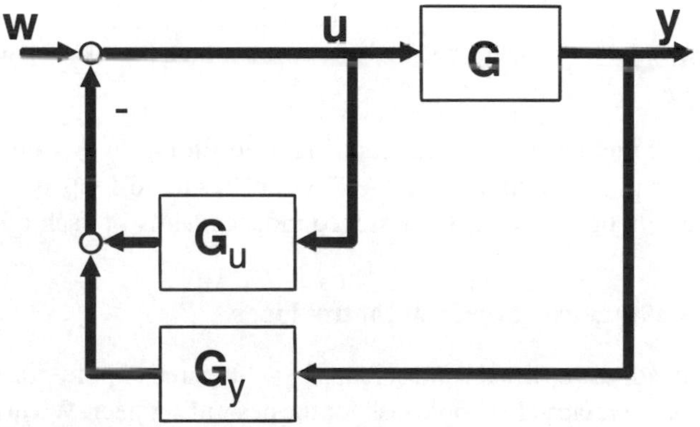

Figure 1.14. Controller/observer structure in the frequency domain

between input and output values of the dynamical feedback. Equation (1.122) shows that the feedback variable $\boldsymbol{f}(s)$ is generated by two transfer-function matrices from the manipulated value $\boldsymbol{u}(s)$ and the output value $\boldsymbol{y}(s)$. This can be described by

$$\boldsymbol{f}(s) = \boldsymbol{G}_u(s)\boldsymbol{u}(s) + \boldsymbol{G}_y(s)\boldsymbol{y}(s). \tag{1.123}$$

Thus, the control loop in Fig. 1.14 has input and output behaviour equivalent to the system represented in Fig. 1.13.

The transfer-function matrix $\boldsymbol{G}_{yw}(s)$ for set point behaviour is given, for proportional state feedback, by

$$\boldsymbol{G}_{yw}(s) = \boldsymbol{B}(s\boldsymbol{I} - \boldsymbol{A} - \boldsymbol{B}\boldsymbol{F})^{-1}\boldsymbol{C}, \tag{1.124}$$

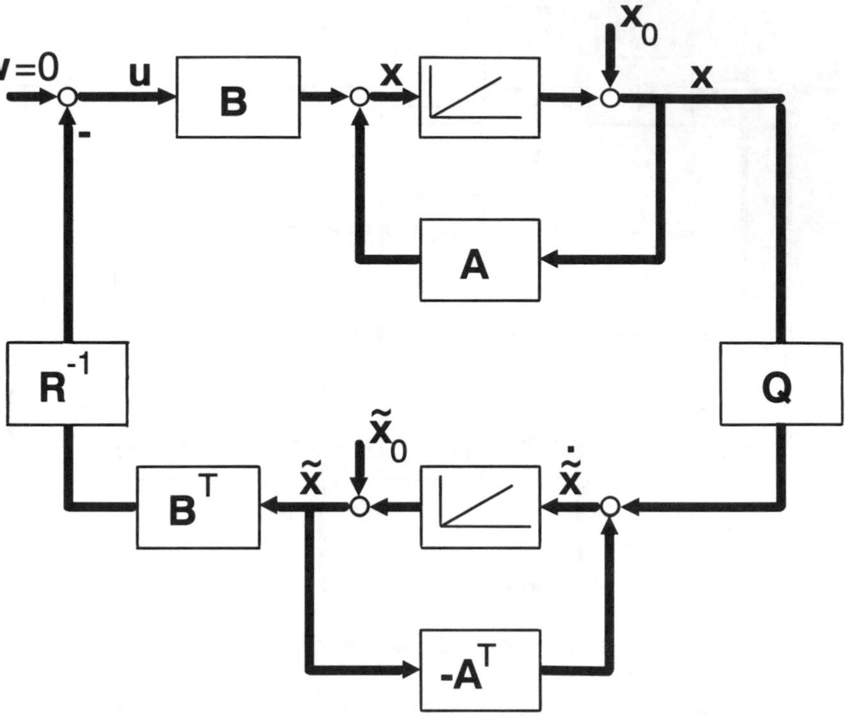

Figure 1.15. Block diagram of the continuous-time canonical system

i.e. the dynamics of the observer do not contribute, according to the separation principle, to the characteristic polynomial $|s\boldsymbol{I} - \boldsymbol{A} - \boldsymbol{B}\boldsymbol{F}|$ of the closed-loop system. This implies that the controller and observer can be designed independently of each other.

1.6.3 Synthesis of quadratic optimal control loops

The synthesis of quadratic optimal control loops is – compared to pole-placement design for non-trivial systems – a relatively simple task for the design engineer. We just have to specify how states, output and input values have to be weighted according to their contribution to the performance criterion. Additionally we can show that the optimal control loop has infinite amplitude margin and a phase margin of at least 60 deg [8].

The continuous-time Hamiltonian system [13] of Fig. 1.15 is considered first, which shows the block diagram associated with the state differential equation

$$\begin{bmatrix} \dot{\boldsymbol{x}}(t) \\ \dot{\boldsymbol{\psi}}(t) \end{bmatrix} = \begin{bmatrix} \boldsymbol{A} & \boldsymbol{B}\boldsymbol{R}^{-1}\boldsymbol{B}^T \\ \boldsymbol{C}^T\boldsymbol{Q} & -\boldsymbol{A}^T \end{bmatrix} \begin{bmatrix} \boldsymbol{x}(t) \\ \boldsymbol{\psi}(t) \end{bmatrix} . \tag{1.125}$$

With the feedback of the state vector x one gets the characteristic polynomial of the system

$$\tilde{\boldsymbol{G}}(s) = [\boldsymbol{I} + \boldsymbol{R}^{-1}\boldsymbol{G}_x^*(s)\boldsymbol{Q}\boldsymbol{G}_x(s)]^{-1} \tag{1.126}$$

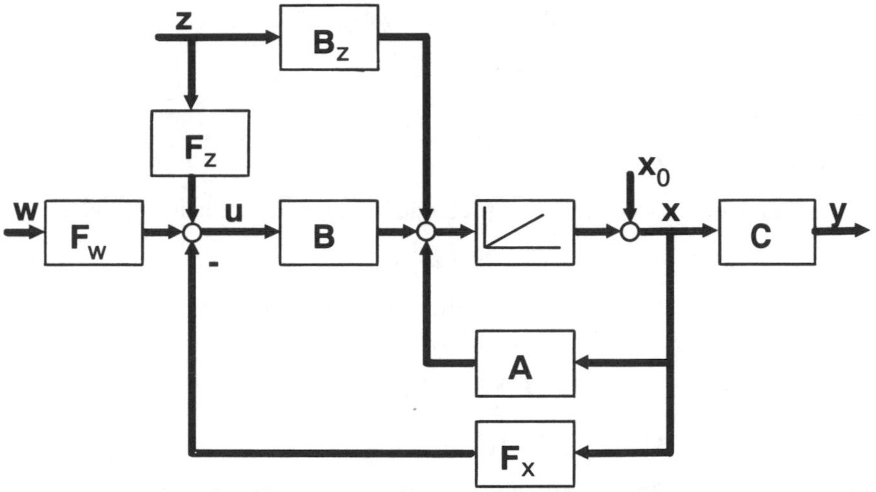

Figure 1.16. Proportional state-feedback structure

which has a symmetric root locus, i.e. the poles of this system are located symmetrically with respect to the imaginary axis of the complex s-plane [8], [11]. A closed-loop system with the stable poles of the canonical system can be generated using quadratic optimal state feedback.

Proportional state feedback

The structure of proportional state feedback is illustrated in Fig. 1.16, and parallels the general structure of Fig. 1.13 except for the feed-forward matrices.

In the course of control synthesis, the performance criterion

$$I_c = \frac{1}{2} \int_0^\infty [\boldsymbol{x}^T(t)\boldsymbol{Q}\boldsymbol{x}(t) + \boldsymbol{u}^T(t)\boldsymbol{R}\boldsymbol{u}(t)]e^{2\alpha t} dt \tag{1.127}$$

will be minimized.

The design engineer can, by choice of the (diagonal) matrices $\boldsymbol{Q}$ and $\boldsymbol{R}$, influence the transient behaviour of the closed-loop system. Large diagonal entries of the positive-definite matrix $\boldsymbol{Q}$ can be specified to penalize deviations of the states from the origin. This will lead to fast regulation of these states; the poles of the closed-loop system will be shifted far to the left in the complex plane. The matrix $\boldsymbol{R}$ can be used to influence the transient behaviour of the control signal $\boldsymbol{u}(t)$. Large diagonal entries r_{ii} lead to smooth transients of this signal and slow regulation of the closed-loop system. If $\boldsymbol{R}$ is chosen as zero (positive semidefinite), the closed-loop system has $n - n_n$ infinite poles and n_n poles situated at the locations of the stable open-loop zeros or unstable zeros mirrored at the imaginary axis. The root locus behaves similarly to the classical root locus method. But in state-feedback design there exist many degrees of freedom, because there is not a single root locus design parameter but two matrices with several entries. Nevertheless, a graphical representation of the root locus with respect to some weighting factors is helpful during design.

The manipulated value is calculated according to

$$\boldsymbol{u}(t) = \boldsymbol{F}_x\boldsymbol{x}(t) + \boldsymbol{F}_z\boldsymbol{z}_s(t) + \boldsymbol{F}_w\boldsymbol{w}(t). \tag{1.128}$$

The state-feedback matrix is

$$\boldsymbol{F}_x = -\boldsymbol{R}^{-1}\boldsymbol{B}^T\boldsymbol{K}, \tag{1.129}$$

where $\boldsymbol{K}$ is the stationary positive-definite solution of the Riccati matrix equation

$$\dot{\boldsymbol{K}} = \boldsymbol{Q} + \boldsymbol{K}\boldsymbol{A} + \boldsymbol{A}^T\boldsymbol{K} - \boldsymbol{K}\boldsymbol{B}\boldsymbol{R}^{-1}\boldsymbol{B}^T\boldsymbol{K}. \tag{1.130}$$

To reach the steady-state set value, the prefilter matrix $\boldsymbol{F}_w$ must compensate the stationary gain of the closed-loop system. This matrix can be evaluated as

$$\begin{aligned}\boldsymbol{F}_w &= [\lim_{s\to 0}\boldsymbol{C}(s\boldsymbol{I} - \boldsymbol{A} - \boldsymbol{B}\boldsymbol{F}_x)^{-1}\boldsymbol{B}]^{-1}\\ &= [\boldsymbol{C}(-\boldsymbol{A} - \boldsymbol{B}\boldsymbol{F}_x)^{-1}\boldsymbol{B}]^{-1}.\end{aligned} \tag{1.131}$$

If the true parameters of the plant do not agree exactly with the design model, then we will not be able to track the set point $\boldsymbol{w}$. As in classical proportional control, there will exist also in the case of proportional state feedback a steady-state control error. Therefore, this type of state feedback is only used where only stabilization is of importance. To get an acceptable reference and disturbance behaviour with vanishing control error, a disturbance observer must be used, or the control structure has to be augmented to proportional-integral state feedback.

Proportional-integral state feedback

Classical control theory shows that the open-loop system must provide an integrator in front of the entry point of the external signal (disturbance, set), to regulate a stepwise change of this signal without steady-state control deviation. This also leads to the introduction of an integral control law in state feedback.

For design of a PI state-feedback controller, the open loop will be augmented by integrators. The system representation is assumed to be a sensor coordinate description, so that the state vector has the form

$$\tilde{\boldsymbol{x}} = \begin{bmatrix}\boldsymbol{x}'\\ \boldsymbol{y}\end{bmatrix}. \tag{1.132}$$

Using

$$\tilde{\boldsymbol{A}} = \begin{bmatrix}\boldsymbol{A}_1 & \boldsymbol{A}_2\\ \boldsymbol{A}_3 & \boldsymbol{A}_4\end{bmatrix}, \quad \tilde{\boldsymbol{B}} = \begin{bmatrix}\boldsymbol{B}_1\\ \boldsymbol{B}_2\end{bmatrix}, \quad \tilde{\boldsymbol{C}} = \begin{bmatrix}\boldsymbol{C}_1 & \boldsymbol{C}_2\end{bmatrix}, \tag{1.133}$$

one generates an augmented system

$$\frac{d\tilde{\boldsymbol{x}}(t)}{dt} = \tilde{\boldsymbol{A}}\tilde{\boldsymbol{x}}(t) + \tilde{\boldsymbol{B}}\boldsymbol{v}(t) \tag{1.134}$$

with

$$\tilde{\boldsymbol{x}}(t) = \begin{bmatrix}\boldsymbol{e}(t)\\ \dot{\boldsymbol{e}}(t)\\ \dot{\boldsymbol{x}}'(t)\end{bmatrix}, \quad \boldsymbol{v}(t) = \dot{\boldsymbol{u}}(t), \quad \boldsymbol{e}(t) = \boldsymbol{w}(t) - \boldsymbol{y}(t) \tag{1.135}$$

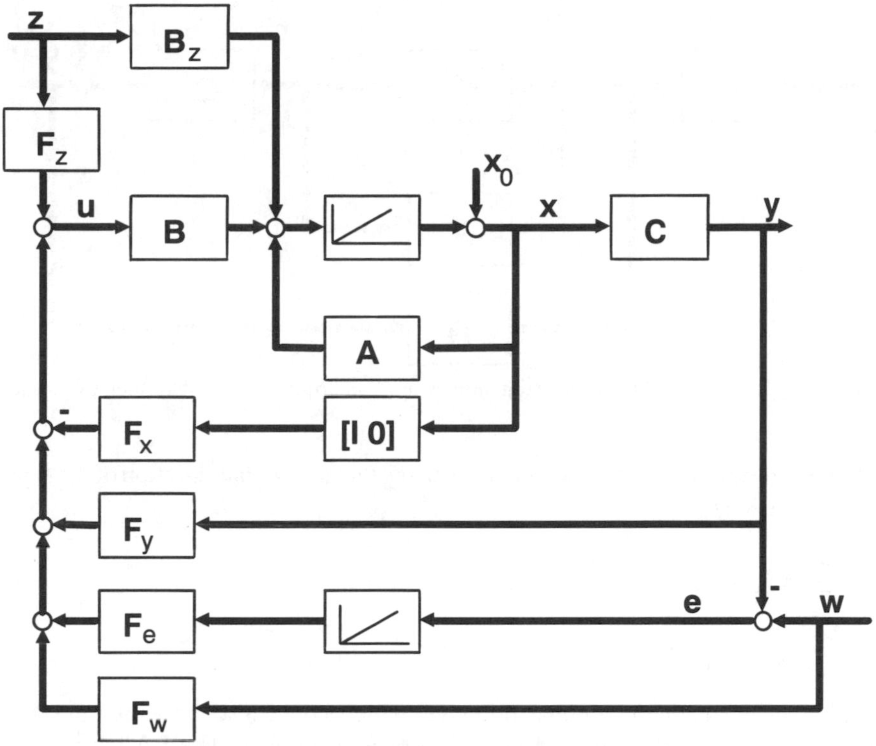

Figure 1.17. Proportional-integral state feedback

and

$$\tilde{A} = \begin{bmatrix} 0 & 0 & 0 \\ I & A_4 & -A_2 \\ 0 & -A_3 & A_1 \end{bmatrix}, \quad \tilde{B} = \begin{bmatrix} 0 \\ -B_2 \\ B_1 \end{bmatrix}. \tag{1.136}$$

Now the performance criterion

$$I_c = \frac{1}{2}\int_0^{\infty} [\tilde{x}^T(t) Q \tilde{x}(t) + v^T(t) \tilde{R} v(t)] e^{2\alpha t} dt \tag{1.137}$$

is used, allowing weighting of the derivative of the control signal and control deviation in addition to weighting of the states. For the choice of the weighting parameters the same guidelines are valid as for a purely proportional control law.

The control signal is evaluated according to Fig. 1.17 with partitioned feedback matrices for states, output values and integrated control error:

$$u(t) = F_{x'} x'(t) + F_y y(t) + F_e \int_0^t e(t)dt + F_z z_s + F_w w(t). \tag{1.138}$$

It should be noted that the set value w does not enter the plant through the integrator, which would lead to sluggish set behaviour, but is fed forward directly to the manipulated value. If the feedback matrix of the augmented control loop is defined as

$$F = [F_e | -F_y | F_x'] = -\tilde{R}^{-1} \tilde{B}^T \tilde{K}, \tag{1.139}$$

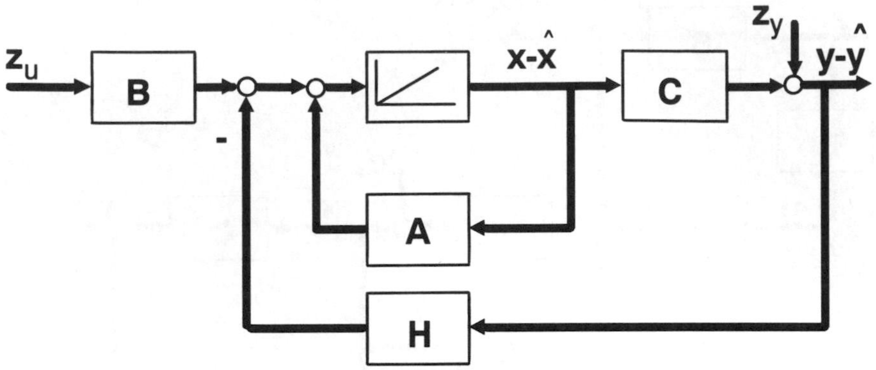

Figure 1.18. Transfer system of the estimation error of an observer with disturbances z_u and z_y

then the controller matrices may be calculated using the steady-state solution of the algebraic Riccati matrix equation

$$\frac{d\tilde{\boldsymbol{K}}}{dt} = \tilde{\boldsymbol{Q}} + \tilde{\boldsymbol{K}}\tilde{\boldsymbol{A}} + \tilde{\boldsymbol{A}}^T\tilde{\boldsymbol{K}} - \tilde{\boldsymbol{K}}\tilde{\boldsymbol{B}}\tilde{\boldsymbol{R}}^{-1}\tilde{\boldsymbol{B}}^T\tilde{\boldsymbol{K}}. \tag{1.140}$$

Design of the observer

The design of a quadratic optimal observer, which could also be called a stationary Kalman–Bucy filter, is dual to the design of a quadratic optimal state-feedback controller. Here, an optimal feedback matrix $\boldsymbol{H}$ must be found, so that the estimation error system as depicted in Fig. 1.18 will be stabilized.

In accordance to Fig. 1.14, we would prefer to structure the observer as in Fig. 1.19. For the full-order observer with disturbance observation, the system has to be augmented by states representing the unmeasurable disturbances. The standard system representation will be substituted by

$$\bar{\boldsymbol{x}}(t) = \begin{bmatrix} \boldsymbol{z}_y \\ \boldsymbol{x} \end{bmatrix}, \quad \bar{\boldsymbol{A}} = \begin{bmatrix} 0 & \boldsymbol{B}_z \\ 0 & \boldsymbol{A} \end{bmatrix}, \quad \bar{\boldsymbol{B}} = \begin{bmatrix} 0 \\ \boldsymbol{B} \end{bmatrix}, \quad \bar{\boldsymbol{C}} = \begin{bmatrix} 0 & \boldsymbol{C} \end{bmatrix}. \tag{1.141}$$

For the matrices of the observer system we can write

$$\boldsymbol{D} = \boldsymbol{A} - \boldsymbol{E}\boldsymbol{C}, \quad \boldsymbol{G} = \boldsymbol{B}, \quad \boldsymbol{G}_z = \boldsymbol{B}_z. \tag{1.142}$$

For the estimation error the performance criterion

$$I_c = \frac{1}{2}\int_0^\infty \{[\boldsymbol{x}(t) - \hat{\boldsymbol{x}}(t)]^T\boldsymbol{Q}[\boldsymbol{x}(t) - \hat{\boldsymbol{x}}(t)] + [\boldsymbol{y}(t) - \hat{\boldsymbol{y}}(t)]^T\boldsymbol{R}[\boldsymbol{y}(t) - \hat{\boldsymbol{y}}(t)]\}e^{2\alpha t}dt \tag{1.143}$$

is minimized.

The feed-forward matrix

$$\boldsymbol{E} = -\boldsymbol{K}\boldsymbol{C}^T\boldsymbol{R}^{-1} \tag{1.144}$$

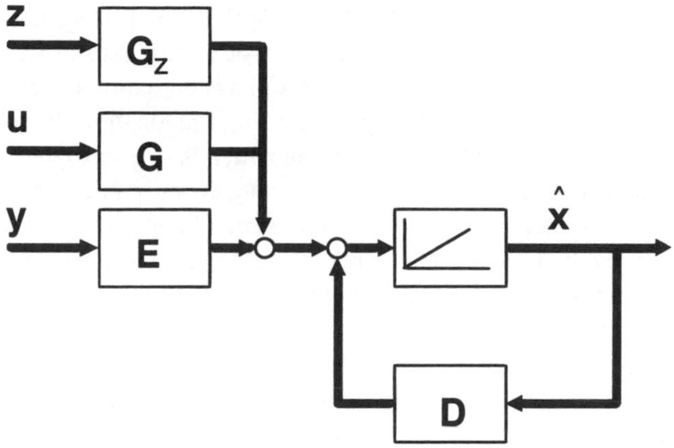

Figure 1.19. Structure of the full-order observer

can be calculated using the positive-definite solution of the algebraic Riccati matrix equation

$$\frac{d\boldsymbol{K}}{dt} = \boldsymbol{Q} + \boldsymbol{K}\boldsymbol{A}^T + \boldsymbol{A}\boldsymbol{K} - \boldsymbol{K}\boldsymbol{C}^T\boldsymbol{R}^{-1}\boldsymbol{C}\boldsymbol{K}. \tag{1.145}$$

This is dual to the design of the proportional state-feedback controller. The meaning of the specific matrices and vectors is as follows:

- $\boldsymbol{D} \in \Re^{n\times n}$: observer system matrix,
- $\boldsymbol{E} \in \Re^{n\times r}$: observer feed-forward matrix for output values,
- $\boldsymbol{G} \in \Re^{n\times m}$: observer feed-forward matrix for plant input values,
- $\boldsymbol{G}_z \in \Re^{n\times l}$: disturbance feed-forward matrix,
- $\hat{\boldsymbol{x}} \in \Re^{n\times 1}$: observed state,
- $\hat{\boldsymbol{y}} \in \Re^{r\times 1}$: reconstructed output.

With the methods discussed here, we are able, using numerical algorithms for pole placement and solution of the Riccati matrix equation, to design state-feedback controllers which assign the dynamics of the closed-loop system for proportional state feedback with an observer and for the augmented control structure with disturbance and reference model, by minimizing a quadratic performance index. In controller synthesis, the design freedom consists mainly in the choice of the weighting matrices. Simulation and graphical display of root loci for variation of certain weights, and evaluation of the maximum and minimum singular values of the return difference matrix, can help the designer.

1.7 CONCLUSIONS

This chapter has presented some fundamentals of system theory and an approach for design of linear control loops. Many important fundamentals of the theory of linear systems have not been mentioned, due to lack of space and to keep this text readable. There exist many good textbooks in this field: [5], [7], [9], [12], [14],[15], [17], [18], [21]. The theory of generalized system matrices introduced by Rosenbrock [19] has not been covered, with its concepts of invariant zeros, decoupling zeros and system equivalence, which can shed more light on observability, controllability and minimal realization. Parts of the textbook of Kailath [8] also cover this field using a notation which is still up to date. Further reading would be worthwhile in this area of system theory using the polynomial matrix approach for design of discrete-time systems [11] and the more general approach by Callier and Desoer [6]. A different view on the fundamental items of linear dynamical systems can be obtained from Wonham [22], who uses a rather abstract geometric approach. For those just wishing to start state-of-the-art controller design for linear systems, the textbook of Åström and Wittenmark [4] is a sufficient introduction to discrete-time systems and linear/quadratic control, especially for those interested in adaptive control. An introduction to stochastic systems can be found in [3].

1.8 LABORATORY EXERCISES

The laboratory session is intended to aid the readers in solving control engineering problems which are discussed in each chapter. The problems discussed in this session are basic in linear control systems and are presented to illustrate the theoretical aspects of control problems given in chapter 1. In this section discussions are limited to linear, time-invariant systems, both continuous and discrete time.

1.8.1 MATLAB software tools applied in laboratory course 1

MATLAB has an excellent collection of commands and functions that are immediately useful for solving control engineering problems. Moreover, SIMULINK as an extension to MATLAB provides a tool for easy simulation of dynamical systems. Hence, SIMULINK is the basic tool that is used in this section. Almost all dynamical systems' models in examples in this section may be either created in the SIMULINK block diagram window or recalled as they were provided as MATLAB m-files. The systems may be further analyzed by entering commands in MATLAB's command window. These commands are included in the Control System Toolbox for MATLAB. For some systems' models, however, it may be simpler to enter them as text in an m-file or so-called MEX-file than to draw the corresponding block diagram. No matter how we create the model, we obtain as a result an S-function. Examples 15, 16, 17 in section 1.8.8 illustrate how to create an S-function for a forced circulation evaporator.

1.8.2 Continuous-time systems: transfer-function and state-space representation

Example 1 – double integrator

Use SIMULINK to check the step response of the double integrator. Build the model using

1. its transfer function,

2. two integrators,

3. its state-space representation.

Quick help

1. Enter the command `simulink` at the MATLAB prompt to open the main block library.

2. Click on the SIMULINK window, and select `New...` from the `File` menu on its menu bar to open a new empty window in which you can construct a system model. The new window is labelled `Untitled`; you can rename it when you save it.

3. Open `Sources` library, and drag `Step Fcn` block into the active window.

4. Open `Sinks` library, and drag `Scope` block into the active window.

5. Open `Linear` library, and drag `Transfer Fcn` block and `Integrator` block into the active window.

6. Open `Connections` library and drag `Inport` and `Outport` blocks into the active window.

7. Place the blocks correctly. After the blocks have been placed, draw lines to connect the blocks by moving the mouse pointer over a block's port and holding down the left mouse button.

8. Open the blocks (by double-clicking) and change some of their internal parameters.

9. Save the system by selecting `Save` from the `File` menu.

10. Run simulation by selecting `Start` from `Simulation` menu.

11. You can adjust one or more of the simulation parameters by selecting `Parameters` from the `Simulation` menu

Verify that you can extract a state-space model from the transfer-function model, by entering the command

```
[A,B,C,D]=linmod('filename')
```

To convert the state-space model to transfer-function form, enter the command

```
[num,den]=ss2tf(A,B,C,D)
```

Questions

1. Are the matrices `A,B,C,D` the same in cases 1 and 2? What is the reason for any differences?

2. Are the transfer functions the same in cases 1 and 2?

Example 2 – sine wave generator

Using SIMULINK, find the model of a signal $x(t)$

$$x(t) = Asin(\omega t + \varphi) \tag{1.146}$$

with parameterization:

$$p = (A, \omega . \varphi)^T \tag{1.147}$$

where:

- $A = 1$,
- $\omega = 1$,
- $\varphi = \pi/2$.

Alternatively, solve the associated differential equation

$$\ddot{x} + x\omega^2 = 0 \tag{1.148}$$

with initial conditions:

$$\dot{x}(0) = f(A, \varphi, \omega), \qquad x(0) = g(A, \varphi)$$

To do this use

- signal generator,
- set of single integrators,
- state-space block.

Observe the system's output. If necessary you can recall file `BD8_2`.

Question

Is the output the same in every solution?

Example 3 – oscillator

Using SIMULINK, build a model of the system that has the transfer function

$$K(s) = \frac{1}{s^2 + 2\xi\omega s + \omega^2} \tag{1.149}$$

where the constants ξ and ω are given the names

- ξ – damping ratio,
- ω – undamped natural frequency.

Sketch the step response of the system under the following conditions

1. underdamped $(0 < \xi)$: $\xi = 0.2; \omega = 1$,
2. critically damped: $\xi = 1; \omega = 1$,
3. stability limit: $\xi = 0; \omega = 1$.

Find the zero-pole model of the system as well as the state-space representation.

Example 4 – DC drive

A DC drive may be described by:

$$u(t) = Ri(t) + L\frac{di(t)}{dt} + E(t) \tag{1.150}$$

$$E(t) = C_e\omega(t) \tag{1.151}$$

$$M_e(t) = C_m i(t) \tag{1.152}$$

$$M_e(t) = M_o + J\frac{d\omega(t)}{dt} \tag{1.153}$$

$$\frac{d\alpha(t)}{dt} = \omega \tag{1.154}$$

where:

- The system's outputs are:

  armature current – $i(t)$,
  speed of rotation – $\omega(t)$,
  angle of shaft rotation – $\alpha(t)$.

- The system's inputs are:

  $u(t) = 500$,
  $M_o(t) = 1000$.

- The system's parameters are:

  $R = 10$,
  $L = 10^{-3}$,
  $C_e = 50$,
  $C_M = 100$,
  $J = 100$.

1. Build a model of the DC drive using single integrators. Simulate the system. Observe the armature current, speed of rotation, and angle of shaft rotation.
2. Extract a state-space model from the single integrator's model. To do this enter the command `[A,B,C,D]=linmod('filename')` at the MATLAB prompt.
3. Convert the state-space model to transfer-function form, entering the command `[num,den]=ss2tf(A,B,C,D)` at the MATLAB prompt.
4. Transfer Eqs (1.150) – (1.154) into a state-space equation. Repeat the simulation using a single state-space block.
5. Convert Eqs (1.150) – (1.154) into a transfer function. Repeat the simulation using the transfer-function block.

If necessary you can recall file `BD8_4`.

Questions

1. How does $u(t)$ influence the system's outputs?
2. How do the system's outputs depend on $M_o(t)$?
3. Compare the state-space models obtained in items 2 and 4. Explain any differences.
4. Compare the transfer-functions models obtained in items 3 and 5.

Example 5 – modelling of a grinding process

The grinding process of a copper concentrator consists of a ball mill, an autogenous mill and a classifier as shown in Fig. 1.20. The ball mill is working in an open loop and the

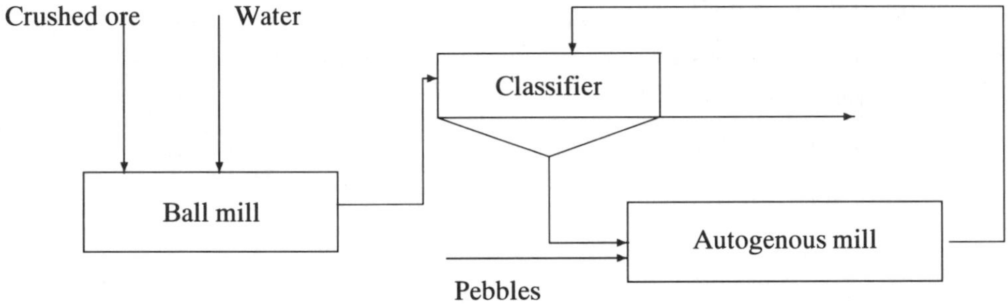

Figure 1.20. The flow diagram of the grinding process

autogenous mill in a closed loop. The main part of the ground ore is fed to the ball mill. Water is added to the feed flow as needed, so that in the output of the mill the solid material concentration is *ca.* 0.75. The pulp flow is directed to the classifier, which separates small particles to the flotation process. The underflow is then pumped to the autogenous mill. The feed of the autogenous mill consists of this underflow and ore pebbles. The output flow is led into the same classifier as the output from the ball mill.

The mathematical model of the grinding process of the copper concentrator [10] is

$$V_1\frac{dx_1}{dt} = u_1w_1 - u_1x_1 \tag{1.155}$$

$$V_1\frac{dx_2}{dt} = u_1x_1 - u_1x_2 \tag{1.156}$$

$$V_2\frac{dx_3}{dt} = k_1u_1x_2 - u_2w_2 + \left\{\frac{k_2k_1u_1}{1-k_2} + \frac{k_2u_2}{1-k_2}\right\}x_4 \tag{1.157}$$

$$- \left\{\frac{k_1u_1}{1-k_2} + \frac{u_2}{1-k_2}\right\}x_3 \tag{1.158}$$

$$V_2\frac{dx_4}{dt} = \left\{\frac{k_1u_1}{1-k_2} + \frac{k_2u_2}{1-k_2}\right\}x_3 - \left\{\frac{k_1u_1}{1-k_2} + \frac{u_2}{1-k_2}\right\}x_4 \tag{1.159}$$

where the parameters of the system are:

$k_1 = 0.8,$

$k_2 = 0.72,$

$\frac{V_1}{u_1} = 7.66 min,$

$$\frac{V_2}{\frac{k_1 u_1}{1-k_2} + \frac{u_2}{1-k_2}} = 3.5 min,$$

$u_1 = 0.9 \frac{t}{min},$

$u_2 = 0.03 \frac{t}{min}.$

The symbols used are:

u_1 – the crush feed,

u_2 – the pebble load,

w_1 – the copper concentration of the crusher feed,

w_2 – the copper concentration of the pebbles,

x_i (i=1,2,3,4) – copper concentrations,

k_1 – the recycle coefficient of the product of the ball mill,

k_2 – the recycle coefficient of the product of the autogenous mill.

Using SIMULINK, build the model of the grinding process. Observe how the process outputs depend on the process input. If necessary you can recall file `BD8_5`.

1.8.3 Discrete-time systems.

Example 6 – simple discrete-time system – time responses

Using SIMULINK, build a discrete-time model of the following continuous-time system:

$$K(s) = \frac{1}{(1+3s)(1+5s)(1+10s)} \tag{1.160}$$

for sampling interval $T_i = 0.5,\ 2,\ 5.$ Compare discrete- and continuous-time step responses. If necessary you can recall file `BD8_6`.

Questions

1. How does the discrete-time step response change with decrease of the sampling interval?

2. How do the discrete-time system's parameters depend on the sampling interval?

Example 7 – simple discrete-time system – pole location

Using SIMULINK, build a model of the following continuous-time system:

$$K(s) = \frac{1}{s^5 + 8s^4 + 19.5s^3 + 19s^2 + 7.5s + 1} \tag{1.161}$$

Find the poles of the continuous-time system. Build a discrete-time model of the continuous-time system for sampling interval $T_i = 0.01, 0.1, 0.5, 1, 2, 5$. Find the poles of discrete-time systems. If necessary you can recall file `BD8_7`.

Question

1. How do the poles change with the sampling interval?

1.8.4 System stability

Example 8 – real-pole feedback system

Using SIMULINK, build a model of the following closed-loop system:

$$\begin{aligned} U(s) &= W(s) - Y(s) \\ V(s) &= kU(s) \\ Y(s) &= K(s)V(s) \end{aligned} \tag{1.162}$$

where:

$$K(s) = \frac{1}{s^2 + 3s + 1}$$

Find the zero-pole representation of the system. Observe the relation between poles of the system, the system's step response and the system's stability for $k = 0.5$, 1 and -1.5. If necessary you can recall file `BD8_8`.

Questions

1. What is the relation between the position of the poles of the system in the s-plane and system stability?
2. What is the relation between the speed of response of the system and the position of the poles of the system in the s-plane?

Example 9 – relation between poles and speed of response of a system

There are four dynamical systems characterized by the set of poles:

1. $p_1 = -0.1, \quad p_2 = -0.5, \quad p_3 = -1,$
2. $p_1 = -0.5, \quad p_2 = -0.5, \quad p_3 = -1,$
3. $p_1 = -1.5, \quad p_2 = -2.0, \quad p_3 = -1,$

4. $p_1 = -0.1, \quad p_2 = -2.0, \quad p_3 = -1.$

Put the following examples in increasing order according to the speed of response of the system.

Example 10 – complex-pole feedback system

Using SIMULINK, build a model of the closed-loop system:

$$\begin{aligned} U(s) &= W(s) - Y(s) \\ V(s) &= kU(s) \\ Y(s) &= K(s)V(s) \end{aligned} \tag{1.163}$$

where:

$$K(s) = \frac{1}{s^2 + 2s + 0.25}$$

Observe the step response of the system. Using MATLAB commands, find the zero-pole model of the system. Observe step responses of several systems with the same real part α of the poles, and different imaginary part, $\pm\beta$. Increase the real part of the poles and repeat the simulation. Observe the damping ratio defined by:

$$r = \frac{A_1}{A_2}$$

where A_1 and A_2 are successive peak deviations of the output from the steady-state value.

Quick help

1. Notice that the transfer function of the above system is $K(s) = \dfrac{1}{(s+\alpha)^2 + \beta^2}$
2. Choose $\alpha = 1$ and $\beta = 0.5;\ 2;\ 5$
3. Repeat simulation for $\alpha = 2$ and $\beta = 0.5;\ 2;\ 5$

Questions

1. How does the damping ratio depend on the position of the poles of the system?
2. How would you wish to design the dynamical system (pole positions) in order to diminish the number of significant oscillations in the transient period?

1.8.5 A design technique based on the elementary methods of stability analysis

Example 11 – stability

For the system:

$$\begin{aligned} U(s) &= W(s) - Y(s) \\ V(s) &= kU(s) \\ Y(s) &= K(s)V(s) \end{aligned} \tag{1.164}$$

where:

$$K(s) = \frac{1}{(1+s)^3}$$

determine that value of gain k which makes all closed-loop poles have a time constant less than 2 sec.

Quick help

1. Notice that the original characteristic equation is

$$s^3 + 3s^2 + 3s + k + 1 = 0$$

2. Carry out the transformation of variables:

$$s = s_0 - \frac{1}{T} = s_0 - \frac{1}{2} \tag{1.165}$$

3. Apply one of the methods of stability analysis to Eq. (1.165) to determine whether all of the closed-loop poles are in the left half of the s-plane.

4. Find the gain k of the system for the feedback system to be stable.

5. Use SIMULINK to observe the system behaviour for several values of the gain k.

1.8.6 System controllability and observability

Example 12 – observability

Check the observability of the system

$$\begin{aligned} \dot{x}_1 &= 2x_1 + x_2 \\ \dot{x}_2 &= -3x_1 - 2x_2 \\ y &= x_1 + x_2 \end{aligned} \tag{1.166}$$

Quick help

1. Use SIMULINK at the MATLAB prompt to build the model

2. At the MATLAB prompt, enter the following commands:

```
[t,x,y]=linsim('filename',10)
plot(t,x)
plot(t,y)
```

Observe the output and states of the system.

3. Use the following commands at the MATLAB prompt to check if the system is observable:

```
[A,B,C,D]=linmod('filename')
Ob=obsv(A,C)
rank(Ob)
```

4. Check the number of unobservable states using the MATLAB command

```
unob=length(A)-rank(Ob)
```

Example 13 – controllability

Check the controllability of the system

$$\begin{aligned}\dot{x}_1 &= 2x_1 + x_2 - u \\ \dot{x}_2 &= -3x_1 - 2x_2 + 3u \\ y &= x_1 + x_2\end{aligned} \tag{1.167}$$

Quick help

1. Use SIMULINK at the MATLAB prompt to build the model.
2. Enter MATLAB commands to check the observability and controllability matrices.
3. Observe the step response as well as the responses of the state variables.

Questions

1. Are both states variables stable?
2. Is the output stable?
3. Are the following conclusions true?

   The output may be stable even though an uncontrollable mode is present. This would be a highly undesirable effect in the operation of a feedback system, since the system output would give no indication of the instability in the feedback information to the controller.

   Even though the controller is receiving information about the instability, it is unable to take corrective action because any regulation of the variable u(t) will have no influence on the instability.

1.8.7 Controllability and observability in feedback system design

Example 14 – pole-zero cancellation

Consider the following system:

$$\begin{aligned}U(s) &= W(s) - Y(s) \\ V(s) &= K_1(s)U(s) \\ Y(s) &= K_2(s)V(s)\end{aligned} \tag{1.168}$$

where:

$$K_1(s) = 2\frac{s+1}{(s+5)(s+2)}$$

$$K_2(s) = \frac{1}{s+1}$$

Explain if the existence of an unobservable or uncontrollable pole in the left half of the s-plane is tolerable.

Quick Help

1. For this system, because of pole-zero cancellation the characteristic equation is

   $$1 + \frac{2}{(s+5)(s+2)}$$

   or

   $$s^2 + 7s + 12 = (s+4)(s+3)$$

   The closed-loop time constants are $1/4$ and $1/3$ sec. Notice that the plant has an open-loop time constant of 0.1 sec. Pole-zero cancellation arouses one's suspicion.

2. Use SIMULINK to build the model of the system.

3. Find a state-space representation.

4. Check the controllability and observability of the system using the following MATLAB commands:

```
[t,x,y]=linsim('name',10);
plot(t,y)
plot(t,x)
[A,B,C,D]=linmod('name')
p=poly(A) (characteristic equation)
r= roots(p) (roots of characteristic equation)
ob=obsv[A,C]
con=cntr[A,B]
unob=length(A)-rank(ob)
uncon=length(A)-rank(con)
```

5. Repeat the experiment for the disturbed system:

$$\begin{aligned} U(s) &= W(s) - Y(s) \\ V(s) &= K_1(s)U(s) \\ Y(s) &= K_2(s)(V(s) + D(s)) \end{aligned} \tag{1.169}$$

Question

What can pole-zero cancellation result in?

1.8.8 Using an S-function for simulation of complex systems

Any non-linear dynamical model can be simulated in SIMULINK by using "standard" blocks incorporated in the library. The user need only collect icons from the library in the user window and connect them. However, even in a simple case the work is boring. It takes a lot of time. The final result can be complicated because all functions e.g. sin, exp, multiplications, additions, integrators, have to be represented by individual blocks. MATLAB-functions allow static mappings between chosen signals. The simulation diagram could be thus simplified; if we agree to hide certain part of the diagram, why not describe all of them in one block, including static mappings as well as model dynamics (mixed, discrete and continuous)? This can be done by using the S-function. In what is follows an example of an S-function describing a forced-circulation evaporator is given. The system is continuous. Another example of S-function for a discrete-time system is given in section 4.8.4.

Example 15 – forced circulation evaporator

Evaporation is one of the basic processes in the chemical industry, where concentration of dilute liquors is demanded. A detailed description of the model is given in [16]. The evaporator variables are as follows:

Variable	Description	Value	Units
F1	feed flowrate	10.0	kg/min
F2	product flowrate	2.0	kg/min
F3	circulating flowrate	50.0	kg/min
F4	vapour flowrate	8.0	kg/min
F5	condensate flowrate	8.0	kg/min
X1	feed composition	5.0	percent
X2	product composition	25.0	percent
T1	feed temperature	40.0	deg C
T2	product temperature	84.6	deg C
T3	vapour temperature	80.6	deg C
L2	separator level	1.0	m
P2	operating pressure	50.5	kPa
F100	steam flowrate	9.3	kg/min
T100	steam temperature	119.9	deg C
P100	steam pressure	194.7	kPa
Q100	heater duty	339.0	kW
F200	cooling water flowrate	208.0	kg/min
T200	cooling water inlet temperature	25.0	deg C
T201	cooling water outlet temperature	46.1	deg C
Q200	condenser duty	307.9	kW

All values refer to the steady state of the system.

There are three mass-balance equation giving three state equations:

- mass balance on the total process liquid in the system, represented by the level $L2$ in the separator:

$$\frac{dL2}{dt} = 0.05(F1 - F4 - F2). \tag{1.170}$$

- mass balance on the solute in the process liquid phase, represented by the level $L2$:

$$\frac{dX2}{dt} = 0.05(F1X1 - F2X2). \tag{1.171}$$

- mass balance on the process vapour, represented by the level $L2$:

$$\frac{dP2}{dt} = 0.25(F4 - F5). \tag{1.172}$$

The other equations are algebraic:

$$T2 = 0.5616P2 + 0.3126X2 + 48.43 \tag{1.173}$$

$$T3 = 0.507P2 + 55.0 \tag{1.174}$$

$$T100 = 0.1538P100 + 90.0 \tag{1.175}$$

$$Q100 = 0.16(F1 + F3)(T100 - T2) \tag{1.176}$$

$$F4 = \frac{Q100 - 0.07F1(T2 - T1)}{38.5} \tag{1.177}$$

$$Q200 = \frac{6.84(T3 - T200)}{1 + \frac{48.86}{F200}} \tag{1.178}$$

$$F5 = \frac{Q200}{38.5} \tag{1.179}$$

One can notice that the system is non-linear. Analysis of degrees of freedom as well as technological requirements lead to the conclusion that the system's states $L2$, $X2$ and $P2$ should be controlled by $F2$, $P100$ and $F200$. Other variables, namely $F3$, $F1$, $X1$, $T1$, and $T200$, are treated as disturbances.

S-function describing the evaporator

The S-function is an element of the SIMULINK environment that is called up according to current simulation time t with different flags. It returns vector sys, which can have different meanings according to the flag. The only difference is the first call, when the S-function also returns initial conditions of the states. Thus, the first line of any S-function is as follows:

```
function [sys, x0] = evaporator(t,x,u,flag,...parameters...)
```

where x means state vector and u is the vector of input variables. For the evaporator,

$$
\begin{aligned}
x &= [L2, X2, P2] \\
u &= [F2, P100, F200] \\
\ldots parameters \ldots &= [F3, F1, X1, T1, T200]
\end{aligned}
$$

The whole S-function is:

```
function [sys, x0] = evaporator(t,x,u,flag,F3,F1,X1,T1,T200)

if flag == 0

  sys = [3, 0, 3, 3, 0, 0]';
  x0 = [1; 25; 50.5] ;

elseif abs(flag) == 1

  T2 = 0.5616 * x(3) + 0.3126 * x(2) + 48.43;
  T3 = 0.507 * x(3) + 55.0;
  T100 = 0.1538 * u(2) + 90.0;
  Q100 = 0.16 * (F1 + F3) * (T100 - T2);
  F4 = (Q100 - 0.07 * F1 * (T2 - T1)) / 38.5;
  Q200 = (6.84 * (T3 - T200)) / (1 + 48.86 / u(3));
  F5 = Q200 / 38.5;

  sys(1) = 0.05 * (F1 - F4 - u(1));
  sys(2) = 0.05 * (F1 * X1  - u(1) * x(2));
  sys(3) = 0.25 * (F4 - F5);

elseif flag ==3

  sys = x;

else

     sys = [];

end
```

The very first call has `flag = 0`. The S-function returns the structure of the system to be simulated (vector `sys`) and initial value of the states (vector `x0`). For the evaporator we have:

`sys(1)` = 3 because there are three continuous-time states,

`sys(2)` = 0 because there are no discrete-time states,

`sys(3)` = 3 because there are three output variables,

`sys(4)` = 3 because there are three input (control or manipulated) variables,

`sys(5)` = 0 because there are no discontinuous roots,

`sys(5)` = 0 because there is no direct path from input to output.

The initial conditions are the steady-state values of the state variables. The call with `flag=0` is performed only once. Then, when simulation proceeds, the S-function is called with flags 1 and 3. `flag = 1` is applied in order to calculate at each step the left-hand-side values in the differential equations, defining increments of the state variables values over one integration step. The width of the step is chosen automatically according to the integration method applied (chosen by the user within the simulation menu). Increments of the state are returned in vector `sys`. Thus, the last three equations within the part referring to `flag=1` are nothing but state equations (1.170), (1.171) and (1.172). Notice that, in the right-hand side of the equations, `u(1)` means `F2` and `x(2)` means `X2`. Other variables are either given as the parameters (`F1`, `X1`) or are calculated in the first set of equations within the part referring to `flag=1` (`F4` and `F5`).

The call with `flag=3` returns the output variable. For the evaporator, the output variables are the state variables.

For any other value of `flag`, vector `sys` is returned empty.

Possible modifications

It is possible to modify the proposed S-function. Here are some examples:

- the initial state can be given from outside the S-function by using additional parameters;
- any local variable, e.g. `T2`, `T3`, `T100`, `Q100`, `Q200`, `F4` and `F5`, or any combination of them, can be put into the output by defining additional states;
- parameters of the system can vary if they are defined as additional input variables. In order to make the simulation faster, we can translate the S-function into C-code and compile it using a specialized compiler (a command `cmex` initializes compilation giving a ready-to-use procedure with extension `mex`). For the evaporator model, the C-version of the S-function is as follows:

```
/* evapor_c.c MEX-file for forced circulating evaporator */

#include <math.h>
#include "matrix.h"
#include "mex.h"

#define NSTATES  3
#define NINPUTS  3
#define NOUTPUTS 3

#define NCOEFFS 5
```

```
/* Number of extra parameters passed in*/
static Matrix *Coeffs[NCOEFFS];
/* Parameters matrix pointer*/
#define pF3    Coeffs[0]
#define pF1    Coeffs[1]
#define pX1    Coeffs[2]
#define pT1    Coeffs[3]
#define pT200  Coeffs[4]

void init_conditions(x0)
double *x0;
{
  x0[0] = 1.0;
  x0[1] = 25;
  x0[2] = 50.5;
}

void derivatives(t,x,u,dx)
double t, *x, *u, *dx;

{

  double T2, T3, T100, Q100, F4, Q200, F5;
  double *F3, *F1, *X1, *T1, *T200;

  F3 = (double *)mxGetPr(pF3);
  F1 = (double *)mxGetPr(pF1);
  X1 = (double *)mxGetPr(pX1);
  T1 = (double *)mxGetPr(pT1);
  T200 = (double *)mxGetPr(pT200);

  T2 = 0.5616 * x[2] + 0.3126 * x[1] + 48.43;
  T3 = 0.507 * x[2] + 55.0;
  T100 = 0.1538 * u[1] + 90.0;
  Q100 = 0.16 * (F1[0] + F3[0]) * (T100 - T2);
  F4 = (Q100 - 0.07 * F1[0] * (T2 - T1[0])) / 38.5;
  Q200 = (6.84 * (T3 - T200[0])) / (1 + 48.86 / u[2]);
  F5 = Q200 / 38.5;

  dx[0] = 0.05 * (F1[0] - F4 - u[0]);
  dx[1] = 0.05 * (F1[0] * X1[0]  - u[0] * x[1]);
  dx[2] = 0.25 * (F4 - F5);

}

void outputs(t,x,u,y)
```

```
double t, *x, *u, *y;
{
  y[0] = x[0];
  y[1] = x[1];
  y[2] = x[2];
}

#include "simulink.h"
```

From the results above it is clear that the following procedures have taken the role of flag

`init_conditions` is responsible for state initialization,

`derivatives` refers to flag=1,

`outputs` refers to flag=3.

The structure of the system is defined by variables `NSTATES`, `NINPUTS` and `NOUTPUTS`. The variable `NCOEFFS` plays a special role. We use it to define matrix `Coeffs`. Elements `pF3`, `pF1`, `pX1`, `pT1` and `pT200` allow us to obtain the current values of the respective parameters passed to the procedure. These parameters are obtained by calling the procedure `mxGetPr`. We should notice the necessity for proper definition of the variables used in all procedures. To put everything into action, one should add `include` commands referring to `<math.h>`, `"matrix.h"` and `"mex.h"` at the beginning and `"simulink.h"` at the end.

Including an S-function in the overall diagram

To incorporate an S-function into overall diagram, the S-function block should be picked up first from the library. Then the name of the S-function as a subsystem function name should be entered, as well as the list of function parameters. We can notice that after icon closing, the S-function name appears within the block. Now, block masking is necessary; we can perform it by choosing the appropriate option in the `Option` menu. The following lines should be filled in:

- `New block type` – the name of the user–type block; here for instance `Non-linear plant`.

- `Dialog strings` – the first string is the local name of the block. After the masking operation, when the block is opened `New block type` and the first `Dialog string` appears within the window and identifies the block; all the subsequent strings appear as the titles of subsequent lines and can be used to identify the parameters of the system. The name of the block is `Forced Circulation evaporator`. Parameters are named as follows: `Circulating flowrate -- F3`, `Feed flowrate -- F1`, `Feed composition -- X1`, `Feed temperature -- T1` and `Cooling water inlet temperature -- T200`.

- `Initialization commands` – initialize the values of the parameters according to the list entered previously. The sequence of initialization does not play a role, but we have to match every parameter with its place in subsequent lines that appear after opening the masked S-function. Thus, `F3 = @1` means that parameter `F3` is entered from the first line.
- `Drawing commands` – should contain the text appearing in the final icon.
- `Help string` – specifies help.

After masking, the final block is a ready-to-use icon. By opening it we can enter all parameters and read help.

Example 16 – evaporator control system I

The SIMULINK function `bd8_9` entered at the MATLAB prompt provides an example of an evaporator control system. By opening the block `EVAPORATOR` we can observe how to enter the S-function parameters. Note that `F1` is not equal to its steady value, given in section 1.8.8; this serves as an example of disturbances. We can unmask the block and observe the whole procedure of incorporating an S-function into the overall diagram (section 1.8.8). The control signal is also provided, either by setting equal its steady values (`P100` and `F200`) or by a control loop with a PID controller (`F2`). The controller stabilizes the level in the separator, thus the reference value is 1.0 m. The second output variable, product composition, can be observed on using a SIMULINK scope block.

Example 17 – evaporator control system II

Try to simulate the system. Change the controller tuning. Add one or two additional controllers constituting loops `X2` – `P100` and/or `P2` – `F200`. Change disturbances (parameters of the S-function). Introduce a MEX-file to check the speed of simulation.

1.9 REFERENCES

[1] K. J. Åström. *Simple Self-tuners.* Lund Report CODEN: LUTF2/(TRFT-7184), Lund Institute, Lund, Sweden, 1979.

[2] K. J. Åström and B. Wittenmark. *Computer Controlled Systems. Theory and Design.* Prentice Hall, Englewood Cliffs, NJ, 1984.

[3] K. J. Åström. *Introduction to Stochastic Control Theory.* Academic Press, New York, 1970.

[4] S. Barnett. *Polynomials and Linear Control Systems.* Marcel Dekker, New York, 1983.

[5] G. Biernson. *Principles of Feedback Control.* Wiley, New York, 1988.

[6] F. M. Callier and C. A. Desoer. *Multivariable Feedback Systems.* Springer, New York, 1982.

[7] D. K. Frederick and A. B. Carlson. *Linear Systems in Communication and Control.* Wiley, New York, 1971.

[8] T. Kailath. *Linear Systems.* Prentice Hall, Englewood Cliffs, NJ, 1980.

[9] E. Kamen. *Introduction to Signals and Systems.* Macmillan, New York, 1987.

[10] U. Kortela and T. Niemi. Modelling and prediction of a grinding process. *Automatica*, 14:547–557, 1978.

[11] V. Kucera. *Discrete Linear Control. The Polynomial Equation Approach.* Wiley, Chichester, 1977.

[12] B. C. Kuo. *Automatic Control Systems.* Prentice Hall, Englewood Cliffs, NJ, 1975.

[13] H. Kwakernaak and R. Sivan. *Linear Optimal Control Systems.* Wiley Interscience, New York, 1972.

[14] B. P. Lathi. *Signals, Systems and Control.* Intext, New York, 1979.

[15] A. G. J. MacFarlane. *Dynamical System Models.* Harrop, London, 1970.

[16] P. L. Lee and R. B. Newell. *Applied Process Control: A Case Study.* Prentice Hall, London, 1989.

[17] K. Ogata. *System Dynamics.* Prentice Hall, Englewood Cliffs, NJ, 1992.

[18] R. V. Patel and N. Munro. *Multivariable System Theory and Design.* Pergamon Press, Oxford, 1982.

[19] H. H. Rosenbrock. *State Space and Multivariable Theory.* Nelson, London, 1970.

[20] J. H. Wilkinson. *The Algebraic Eigenvalue Problem.* Oxford University Press, Oxford, 1965.

[21] W. A. Wolovich. *Multivariable Systems.* Springer, New York, 1973.

[22] W. M. Wonham. *Linear Multivariable Control: A Geometric Approach.* Springer, New York, 1979.

2

Control system design

2.1 INTRODUCTION

The first simulation tools for control system design were based on analog techniques. The rapid development of digital computers has brought new and widely used software tools like MATLAB and SIMULINK. Those tools are equipped with libraries of classical and modern control design methods that free the designer from time-consuming programming. They allow validation of control system behaviour through experiments performed with a mathematical model of the system. The comparison of simulation results with real-life responses of physical systems and the ability to introduce corrections easily at the simulation stage allow control engineers to choose the best solution for a given control system design problem.

The purpose of this chapter is to give some background in the most popular methods from the control engineer's tool set, together with simulation examples. Theoretical background of the transfer function, uncertainty models and useful scalar signal and system measures are given. The ideas of feedback, stability and measures of quality for closed-loop systems are also introduced. Classical control design schemes for both SISO and multivariable plants are presented, together with more involved linear/quadratic control problems and basics of H_∞ control.

2.2 SISO PLANT REPRESENTATION

2.2.1 Transfer function and frequency response

A linear, time-invariant, dynamical system, connecting input signal $u(t)$ with output signal $y(t)$, can be described in the time domain by a linear, constant-coefficient differential equation. The ratio of the Laplace transform of the output signal to the Laplace transform of the input signal, with initial conditions assumed to be zero, is known as the transfer function:

$$K_o(s) = \frac{Y(s)}{U(s)}. \tag{2.1}$$

In general, the transfer function is of the form:

$$K_o(s) = e^{-sT_0}\frac{b_0 + b_1 s + \cdots + b_m s^m}{a_0 + a_1 s + \cdots + s^n}, \tag{2.2}$$

where T_0 is the time delay and $m < n$. The roots of the numerator are the transfer function zeros. If any zero is in the closed right-half s-plane, then the system is called non-minimum phase, otherwise it is minimum phase. Similarly, the roots of the denominator are poles of $K_o(s)$. If any pole is located in the closed right-half s-plane, then the system is unstable; otherwise it is stable. Almost all real physical systems are characterized by proper transfer functions which satisfy:

$$\lim_{s \to \infty} K_o(s) = 0. \tag{2.3}$$

The transfer function is a complex-valued function. Substitution $s = j\omega$ in the transfer function $K_o(s)$ results in a frequency response $K_o(j\omega)$. $K_o(j\omega)$ may be represented in several ways:

- as the Nyquist plot of its imaginary part versus its real part in the complex plane;
- as a pair of plots of the frequency response magnitude and angle versus frequency ω with linear or logarithmic scales of the frequency and amplitude. Special cases are Bode plots in which the logarithm to the base 10 of gain is multiplied by 20 (to give a value in dB) and the frequency is on a logarithmic scale. Minimum phase systems can be uniquely characterized by the frequency response gain;
- as a pair of plots of the frequency response real part versus frequency and imaginary part versus frequency on the linear or logarithmic frequency scale.

The Nyquist and Bode plots of the system:

$$K_o(s) = \frac{2}{(1+s)^3} \tag{2.4}$$

are presented in Figs 2.1 and 2.2, respectively.

2.2.2 Uncertainty

Real-world physical systems cannot be modelled exactly by nominal mathematical relations. Modelling errors should be taken into account because they influence the performance of the control system to be designed. The control system should maintain desired behaviour despite its uncertainty.

Suppose that a system is modelled by $K_o(s)$ and the actual model is $\tilde{K}_o(s)$. Uncertainty models may be divided into two categories:

- the structured uncertainty, which is represented by ranges or bounds on system parameters;
- the unstructured uncertainty, which is given by bounds for the location of the frequency response of the system.

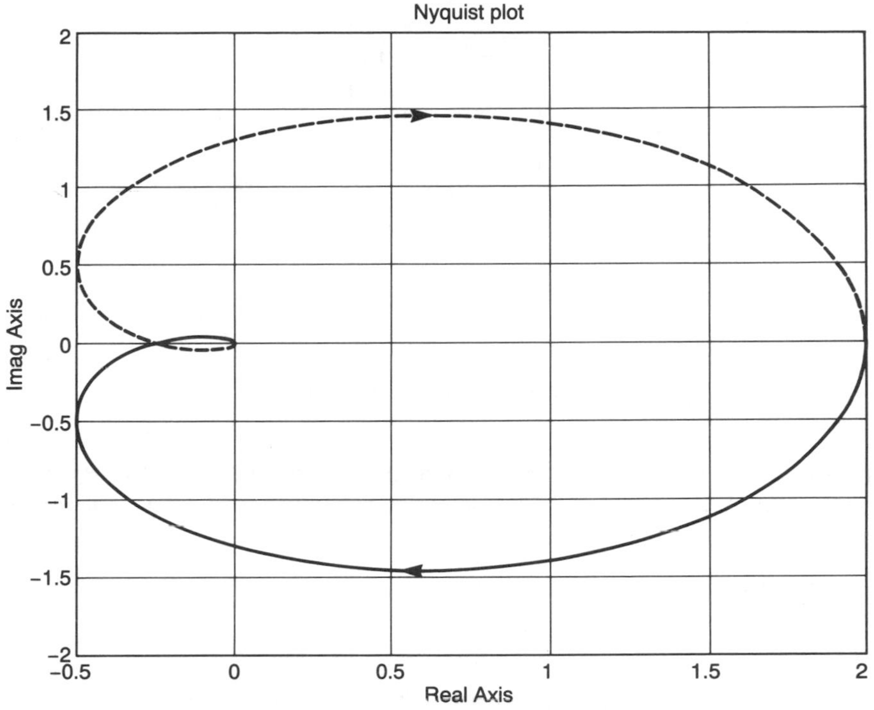

Figure 2.1. Nyquist plot of $K_o(s)$

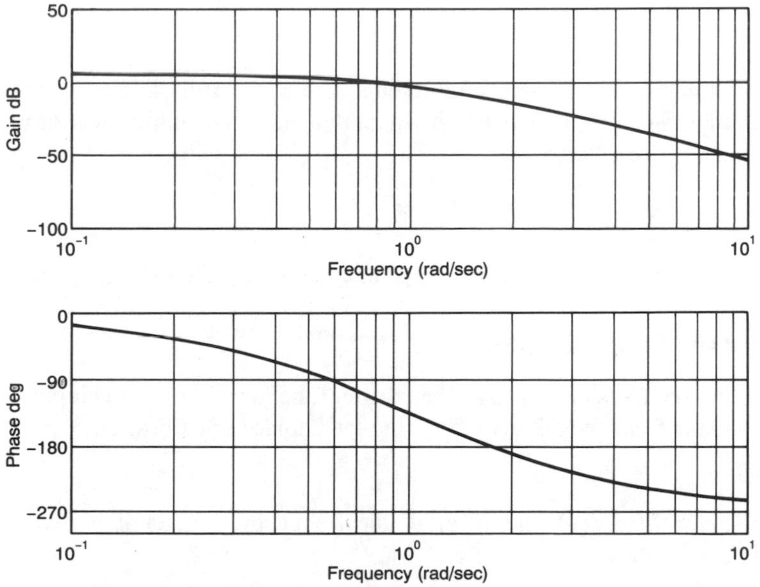

Figure 2.2. Bode plots of $K_o(s)$

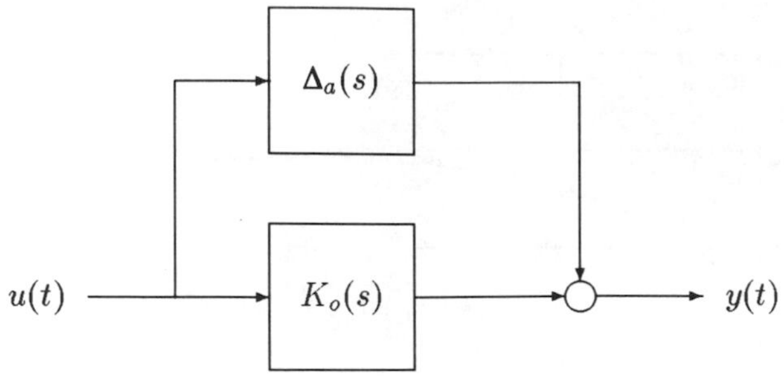

Figure 2.3. Additive uncertainty

The unstructured uncertainty is more important then the structured one because all models used in the control system design should include uncertainty to cover unmodelled dynamics. This uncertainty can be modelled as:

- The additive uncertainty $\Delta_a(s)$ (Fig. 2.3):

$$\Delta_a(s) = \tilde{K}_o(s) - K_o(s). \tag{2.5}$$

  This uncertainty is used to model errors in high-frequency dynamics.

- The multiplicative uncertainty $\Delta_m(s)$:

$$\Delta_m(s) = \frac{\tilde{K}_o(s) - K_o(s)}{K_o(s)}. \tag{2.6}$$

  The multiplicative uncertainty represents a relative error in model. This kind of uncertainty can be applied to model sensor dynamics. The multiplicative uncertainty can be also represented as:

$$\tilde{K}_o(s) = \left(1 + \Delta_m(s)\right) K_o(s). \tag{2.7}$$

These two representations of the unstructured uncertainty are the most frequently used.

2.2.3 Signal and system measures

Output signal $y(t)$ of a system given by the transfer function $K_o(s)$ to any input signal $u(t)$ can be found in the frequency domain by using the Laplace transform as:

$$Y(s) = K_o(s)U(s). \tag{2.8}$$

The multiplication of $K_o(s)$ and $U(s)$ in the frequency domain is equivalent to the time-domain convolution:

$$y(t) = \int_0^t k(\tau)u(t-\tau)d\tau, \tag{2.9}$$

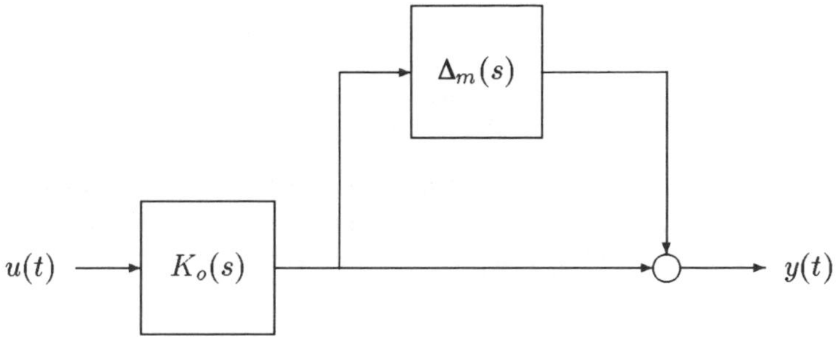

Figure 2.4. Multiplicative uncertainty

where $k(t)$ is the impulse response of the system. The performance of this system can be measured by a size of signals $u(t)$, $y(t)$ and a size of the transfer function $K_o(s)$, and its uncertainty.

An important signal characterization is a measure of its magnitude. The L_p norm of the signal $u(t)$ is

$$||u(t)||_p = \left(\int_{-\infty}^{\infty} |u(t)|^p dt\right)^{\frac{1}{p}} \tag{2.10}$$

where p is a positive integer. It represents a single non-negative number that provides an overall measure of the signal size. The following three norms have particular significance in the control system design:

- the L_1-norm:

$$||u(t)||_1 = \int_{-\infty}^{\infty} |u(t)|dt; \tag{2.11}$$

- the L_2-norm, which provides a measure of the signal power:

$$||u(t)||_2 = \sqrt{\int_{-\infty}^{\infty} (u(t))^2 dt}; \tag{2.12}$$

- the L_∞-norm, which is the least upper bound on the signal absolute value:

$$||u(t)||_\infty = \sup_t |u(t)|. \tag{2.13}$$

The size of the transfer function $K_o(s)$ and the system uncertainty can be measured by using the following norms:

- The H_2-norm:

$$\|K_o(j\omega)\|_2 = \sqrt{\frac{1}{2\pi}\int_{-\infty}^{\infty} |K_o(j\omega)|^2 \, d\omega} \tag{2.14}$$

This norm can be interpreted as the mean square value of the output signal when the system is driven by white noise. The H_2-norm can be calculated for a strictly proper transfer function $K_o(s)$ by using the spectral factorization and residue theorems:

$$\|K_o(j\omega)\|_2 = \sqrt{\frac{1}{2\pi j}\oint K_o(s)K_o(-s)ds}. \tag{2.15}$$

- The H_∞-norm:

$$\|K_o(j\omega)\|_\infty = \sup_{u(t)\neq 0} \frac{\|y(t)\|_2}{\|u(t)\|_2}, \tag{2.16}$$

where $y(t)$ and $u(t)$ are the system output and input signals, respectively. For stable systems this definition has the form:

$$\|K_o(j\omega)\|_\infty = \sup_{\omega} |K_o(j\omega)|. \tag{2.17}$$

The value of this norm corresponds to the peak on the magnitude Bode plot for the system. The H_∞ norm satisfies the following sub-multiplicative property (important from the control system design point of view):

$$\|K_o(j\omega)K_r(j\omega)\|_\infty \leq \|K_o(j\omega)\|_\infty \|K_r(j\omega)\|_\infty. \tag{2.18}$$

The choice of appropriate norm depends on the control system design problem. The following dependencies are useful:

$$\|y(t)\|_2 \leq \|K_o(j\omega)\|_\infty \|u(t)\|_2, \tag{2.19}$$
$$\|y(t)\|_\infty \leq \|K_o(j\omega)\|_2 \|u(t)\|_2, \tag{2.20}$$
$$\|y(t)\|_\infty \leq \|k(t)\|_1 \|u(t)\|_\infty. \tag{2.21}$$

From above inequalities it follows that only the H_∞-norm can be interpreted as a gain of the plant $K_o(s)$.

2.3 FEEDBACK SYSTEM

2.3.1 Basic concepts

The structure of a negative-feedback control system is represented in Fig. 2.5. The following notation has been used:

$K_o(s)$ – plant transfer function;

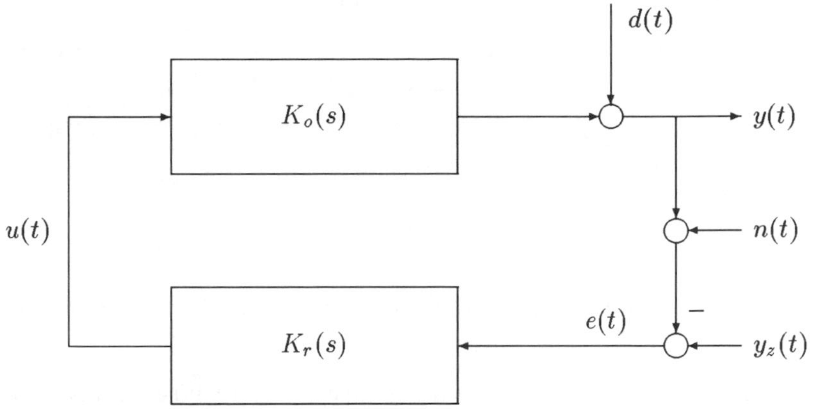

Figure 2.5. Feedback control system

$K_r(s)$ – controller transfer function;

$y(t)$ – output signal;

$y_z(t)$ – reference signal;

$d(t)$ – disturbance at the output;

$u(t)$ – measurement noise;

$e(t)$ – tracking error.

If the Laplace transforms of signals $y_z(t)$, $d(t)$ and $d(t)$ exist, the transform of the output signal of the closed-loop system is:

$$Y(s) = \frac{K_r(s)K_o(s)}{1 + K_r(s)K_o(s)}\left(Y_z(s) - N(s)\right) + \frac{1}{1 + K_r(s)K_o(s)}D(s). \tag{2.22}$$

In this equation, the product $K_r(s)K_o(s)$ is called the open-loop transfer function. The factor

$$M(s) = \frac{K_r(s)K_o(s)}{1 + K_r(s)K_o(s)}, \tag{2.23}$$

which represents the influence of $Y_z(s)$ on $Y(s)$, is known as the complementary sensitivity function or the closed-loop transfer function. The second factor

$$S(s) = \frac{1}{1 + K_r(s)K_o(s)}, \tag{2.24}$$

which is the measure of disturbance rejection, is called the sensitivity function. In general the control system should follow the reference signal $Y_z(s)$ with sufficiently small error, in order to reduce the influence of disturbances and measurement noise, i.e.:

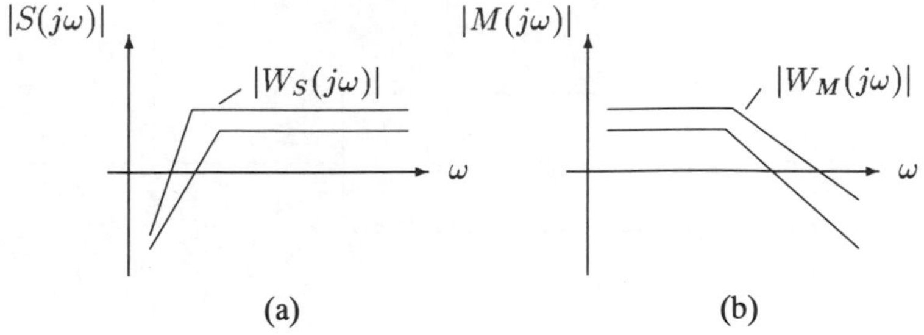

Figure 2.6. Sensitivity (a) and complementary sensitivity (b) functions and their bounds

- to reduce the influence of disturbances, the sensitivity function should be small for frequencies present in the disturbances;
- to reduce the influence of measurement noise, the complementary sensitivity function should be small for frequencies present in this noise;
- to reduce the tracking error, the complementary sensitivity function should be close to 1 for frequencies present in the reference signal.

For large gains ($|K_r(s)K_o(s)| \gg 1$) the sensitivity function approaches 0 and the complementary sensitivity function tends to 1. Therefore the tracking and disturbance rejection are compatible requirements. They conflict with the measurement noise suppression.

These specifications can be converted into frequency-dependent bounds on the sensitivity and complementary sensitivity functions (Fig. 2.6). They are usually approximated by gains of transfer functions $W_S(s)$ and $W_M(s)$:

$$|S(j\omega)| \leq |W_S(j\omega)|, \tag{2.25}$$
$$|M(j\omega)| \leq |W_M(j\omega)|. \tag{2.26}$$

for all frequencies ω. These transfer functions are chosen in an arbitrary manner.

Norms provide a means of converting these frequency-dependent specifications into single numbers. The sensitivity and complementary sensitivity functions performance are stated as:

$$||W_S^{-1}(j\omega)S(j\omega)||_\infty \leq 1, \tag{2.27}$$
$$||W_M^{-1}(j\omega)M(j\omega)||_\infty \leq 1. \tag{2.28}$$

Bounds on the sensitivity and complementary sensitivity functions can be recalculated into bounds that should be met by the gain of the open-loop transfer function (Fig. 2.7).

There are some very interesting interpretations of the sensitivity function. The first one presented here follows from another definition of the sensitivity function:

$$|S(j\omega)| = \left|\frac{E(j\omega)}{E_o(j\omega)}\right|, \tag{2.29}$$

where:

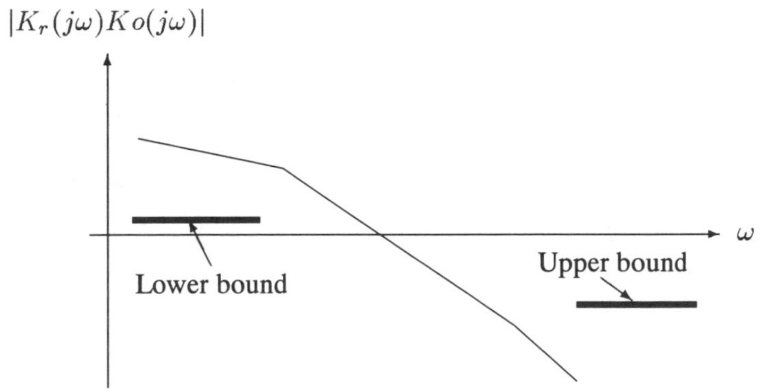

Figure 2.7. Bounds on the open-loop transfer-function gain

$E(j\omega)$ – the closed-loop tracking error;

$E_o(j\omega)$ – the open-loop tracking error.

Thus $|S(j\omega)|$ gives information about the ratio of the tracking error in the closed-loop system to the tracking error in the open-loop system. Disturbances for frequencies ω such that $|S(\omega)| < 1$ can be reduced by the closed-loop system. It can be proved that for $|S(j\omega)| < 1$ the closed-loop system is less sensitive to parameter variations of the plant than the open-loop system with the same transfer function when:

$$\int_0^t e^2(t) < \int_0^t e_o^2(t). \tag{2.30}$$

The overall system is still very sensitive to variations of controller parameters, and therefore it is possible to tune the system behaviour by changing the controller parameters.

The closed-loop system accuracy may be represented by the steady-state error:

$$\lim_{t\to\infty} e(t) - \lim_{t\to\infty} (y_s(t) - y(t)). \tag{2.31}$$

This error depends on the structure of the open-loop transfer function and the properties of the reference signal. If $y_z(t)$ is a step signal ($Y_z(s) = \frac{1}{s}$), then

$$\lim_{t\to\infty} e(t) = \lim_{s\to 0} \frac{1}{1 + K_r(s)K_o(s)}. \tag{2.32}$$

There exist two additional design criteria: overshoot and settling time. They are closely connected with the step response $h(t)$ of the closed-loop system. The overshoot of an asymptotically stable system is defined as:

$$\eta = \frac{\max_{t>0} h(t) - h(\infty)}{h(\infty)} 100\%. \tag{2.33}$$

The settling time t_s is the time at which the step response reaches the steady-state value with an accuracy of 5%. The initial speed of the step response defined as the time it takes this response to go from 10% to 90% of its steady-state value is known as the rise time.

2.3.2 Stability

The stability of the system depends on the location of its poles. If all poles of the closed-loop transfer function are in the left-half s-plane, then the closed-loop system is asymptotically stable. If one or more poles are in the right-half s-plane, then the closed-loop system is unstable. The closed-loop stability can be checked by calculation of the poles of the closed-loop transfer function. Many problems arise for higher-order systems, so it is worth using methods which do not involve calculating poles.

The characteristic equation for the closed-loop system is:

$$1 + K_r(s)K_o(s) = 0. \tag{2.34}$$

For systems without time delay the characteristic equation can be shown to be of the following form:

$$\alpha_n s^n + \alpha_{n-1}s^{n-1} + \cdots + \alpha_1 s + \alpha_0 = 0. \tag{2.35}$$

The closed-loop system is stable if all roots of this equation are found to be in the left-half s-plane. It is equivalent to the following necessary and sufficient conditions:

- all coefficients of the characteristic equation are greater than zero:

$$\alpha_0 > 0, \alpha_1 > 0, \ldots, \alpha_n > 0; \tag{2.36}$$

- the matrix

$$\begin{bmatrix} \alpha_{n-1} & \alpha_{n-3} & \alpha_{n-5} & \ldots & 0 \\ \alpha_n & \alpha_{n-2} & \alpha_{n-4} & \ldots & 0 \\ 0 & \alpha_{n-1} & \alpha_{n-3} & \ldots & 0 \\ 0 & \alpha_{n-2} & \alpha_{n-4} & \ldots & 0 \\ \vdots & \vdots & \vdots & \vdots & 0 \\ 0 & 0 & 0 & \ldots & \alpha_0 \end{bmatrix} \tag{2.37}$$

  is strictly positive definite.

The above, so-called Hurwitz stability criterion can be effectively used only for systems without time delay. When the closed-loop system is unstable, the Hurwitz test does not allow us to estimate the number of the poles in the right-half s-plane and there emerges a necessity of using, for example, the Routh array method.

The Routh array is defined as an n-row table:

$$\begin{vmatrix} \alpha_n & \alpha_{n-2} & \alpha_{n-4} & \ldots \\ \alpha_{n-1} & \alpha_{n-3} & \alpha_{n-5} & \ldots \\ d_{3,1} & d_{3,2} & d_{3,3} & \ldots \\ \vdots & \vdots & \vdots & \vdots \end{vmatrix}, \tag{2.38}$$

where:

$$d_{i,j} = \frac{\begin{vmatrix} d_{(i-2),1} & d_{(i-2),(j+1)} \\ d_{(i-1),1} & d_{(i-1),(j+1)} \end{vmatrix}}{-d_{i-1,1}}. \tag{2.39}$$

The closed-loop system is stable if all first-column coefficients are different from zero and do not change sign. The number of the poles in the right-half s-plane is equal to the number of sign changes in the first-column.

Systems with time delay can be handled using their frequency response. The Nyquist plot of the open-loop transfer function provides a tool for determining stability known as the Nyquist stability criterion. The closed-loop system is asymptotically stable if and only if the graph of the open-loop transfer function $K_r(j\omega)K_o(j\omega)$ for frequencies $-\infty < \omega < \infty$ encircles the $-1 + j0$ point as many times anti-clockwise as $K_r(s)K_o(s)$ has right s half-plane poles. As a special case, when the open-loop system is stable, the stability of the closed-loop system is guaranteed if the frequency response does not encircle the $-1 + j0$ point.

The closed-loop transfer function could be stable but the internal signal $e(t)$, $u(t)$ or $y(t)$ could be unbounded, causing damage of the control system. During the control system design process it is important to ensure that all transfer functions between all inputs ($n(t)$, $y_z(t)$ and $d(t)$) and outputs ($e(t)$, $u(t)$ and $y(t)$) are stable. Feedback systems satisfying this requirement are said to be internally stable. This stability guarantees bounded internal signals for all bounded input signals.

A necessary and sufficient condition of the internal stability for the closed-loop system is that there are no closed-loop poles in the right-half s-plane.

2.3.3 Root locus

Suppose that a feedback transfer function is $kK_r(s)$. The corresponding closed-loop transfer function is:

$$M(s) = \frac{kK_r(s)K_o(s)}{1 + kK_r(s)K_o(s)}, \tag{2.40}$$

where k is the feedback gain. It is clear that closed-loop zeros are zeros of the open-loop transfer function $kK_r(s)K_o(s)$, assuming that there are no pole zero cancellations. The location of the poles of the closed-loop transfer function depends on the magnitude of k. The plot of the roots of the characteristic equation

$$1 + kK_r(s)K_o(s) = 0, \tag{2.41}$$

in the complex plane as a function of parameter k varying from 0 to ∞ is known as the root locus.

The root locus aids analysis and design processes of closed-loop systems in checking asymptotic high-gain behaviour for different feedback transfer functions, and evaluating the time and frequency responses.

2.3.4 Robust stability and performance

Consider that a controller $K_r(s)$ stabilizes the nominal plant model $K_o(s)$. Additionally, we require that the controller be designed to ensure stability and meet performance specifications for all possible plants defined by an uncertainty. This behaviour of the closed-loop system is called robust stability and robust performance.

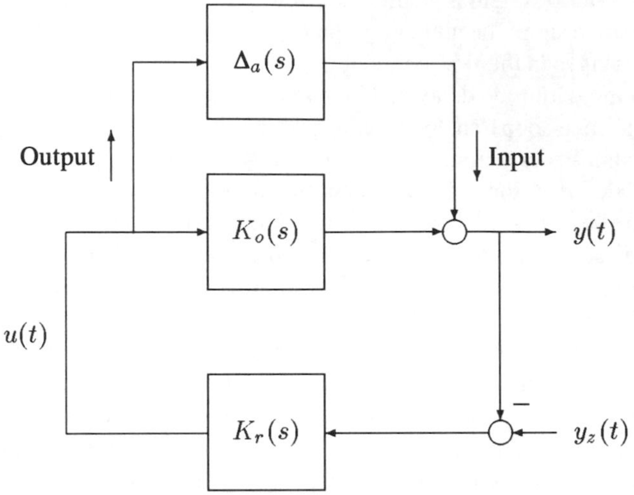

Figure 2.8. Control system with additive uncertainty

Conditions for robust stability can be derived from a modified Nyquist stability criterion known as the small gain theorem. Assuming that the open-loop system is stable, the distance from the critical point $-1 + j0$ to the nearest point on the Nyquist graph of the open-loop transfer function $K_r(s)K_o(s)$ is calculated. The closed-loop system is stable if for all frequencies ω the open-loop gain is kept small:

$$|K_r(j\omega)K_o(j\omega)| < 1 \tag{2.42}$$

Using the H_∞-norm we can express it as:

$$||K_r(j\omega)K_o(j\omega)||_\infty < 1. \tag{2.43}$$

The small gain theorem ensures internal stability. From the sub-multiplicative property it follows that the robust stability of the closed-loop system is also guaranteed if for all ω the following condition holds:

$$|K_r(j\omega)|\,|K_o(j\omega)| < 1. \tag{2.44}$$

For stable additive and multiplicative uncertainty transfer functions $\Delta_a(s)$ and $\Delta_m(s)$ under the small gain theorem, the closed-loop system (Figs 2.8 or 2.9) will remain robustly stable if for all frequencies ω uncertainty models satisfy the following conditions:

- additive uncertainty:

$$|\Delta_a(j\omega)| < \frac{1}{|K_r(j\omega)S(j\omega)|}; \tag{2.45}$$

- multiplicative uncertainty:

$$|\Delta_m(j\omega)| < \frac{1}{|M(j\omega)|}. \tag{2.46}$$

Using the H_∞-norm these inequalities are expressed as:

$$||\Delta_a(j\omega)K_r(j\omega)S(j\omega)||_\infty < 1, \tag{2.47}$$

$$||\Delta_m(j\omega)M(j\omega)||_\infty < 1. \tag{2.48}$$

These conditions allow us to find the size of the smallest stable additive or multiplicative uncertainty that will destabilize the closed-loop system. These sizes are defined as:

- the additive robust stability margin:

$$ASM = \frac{1}{||K_r(j\omega)S(j\omega)||_\infty}; \tag{2.49}$$

- the multiplicative robust stability margin:

$$MSM = \frac{1}{||M(j\omega)||_\infty}. \tag{2.50}$$

The large value of the MSM results in small values of the complementary sensitivity function. This is compatible with measurement noise suppression but conflicts with disturbance rejection and tracking accuracy.

Suppose that uncertainties $||\Delta_a(j\omega)||_\infty$ and $||\Delta_m(j\omega)||_\infty$ are bounded. The closed-loop system (Figs 2.8 or 2.9) will be robustly stable if the corresponding condition is fulfilled:

$$|M(j\omega)| < \frac{1}{||\Delta_a(j\omega)||_\infty}, \tag{2.51}$$

$$|K_r(j\omega)S(j\omega)| < \frac{1}{||\Delta_m(j\omega)||_\infty}. \tag{2.52}$$

When the robust stability condition is fulfilled, closed-loop performance requirements expressed in bounds on the sensitivity or (and) complementary sensitivity functions (2.27) should be met.

The sensitivity function for the plant perturbed by the multiplicative uncertainty is:

$$\frac{1}{1 + (1 + \Delta_m(s))\, K_r(s)K_o(s)} = \frac{S(s)}{1 + \Delta_m(s)M(s)}, \tag{2.53}$$

and the robust performance is:

$$\left\| \frac{W_S^{-1}(j\omega)S(j\omega)}{1 + \Delta_m(j\omega)M(j\omega)} \right\|_\infty < 1. \tag{2.54}$$

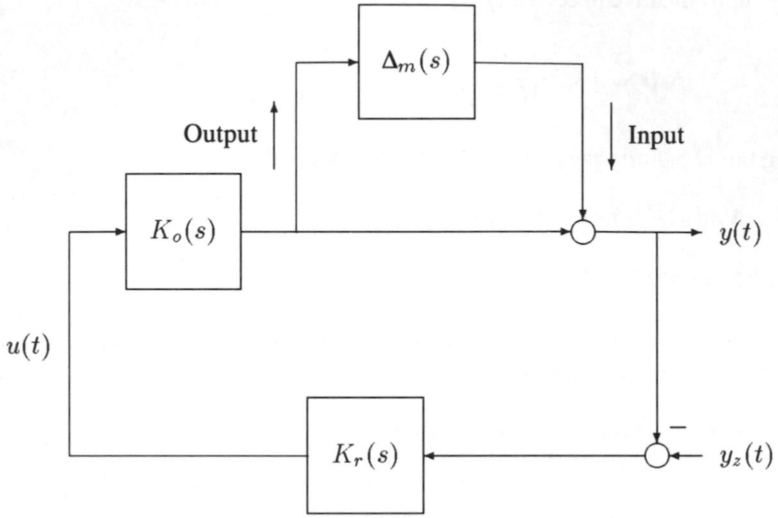

Figure 2.9. Control system with multiplicative uncertainty

A necessary and sufficient condition for simultaneously achieving the robust stability and desired performance of the closed-loop system measured in terms of the sensitivity function is:

$$\left\| \left| W_1^{-1}(j\omega)S(j\omega) \right| + \left| \Delta_m(j\omega)M(j\omega) \right| \right\|_\infty < 1. \tag{2.55}$$

The corresponding condition for the additive uncertainty is of the form:

$$\left\| \left| W_S^{-1}(j\omega)S(j\omega) \right| + \left| \Delta_a(j\omega)K_r(j\omega)S(j\omega) \right| \right\|_\infty < 1. \tag{2.56}$$

Similar tests for bounds on the complementary sensitivity function cannot be expressed in a simple way.

2.4 CLASSICAL DESIGN PRINCIPLES

2.4.1 Stability margin

The system model used by an engineer in control system design is very often biased by uncertainties due to gain and phase changes. For a stable closed-loop system it is necessary to ensure an adequate stability margin as a measure of the distance between frequency response of the open-loop transfer function and the $-1 + j0$ point in the complex plane, known as the gain and phase margins.

The gain margin ΔK is defined for the frequency at which the phase response is equal to $-\pi$. It is the maximum gain increase in dB for which the open-loop frequency response reaches the $-1 + j0$ point. The frequency at which the phase is equal to $-\pi$ is called the phase crossover frequency (see Fig. 2.10).

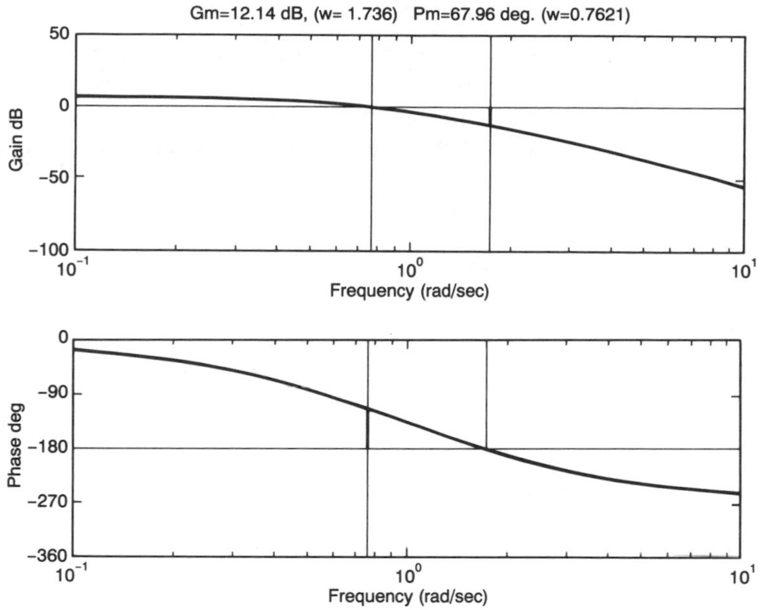

Figure 2.10. Gain and phase margins

The phase margin $\Delta\phi$ is a minimum amount of phase lag that must be added to the system to destabilize it so that the open-loop frequency response reaches the $-1 + j0$ point (see Fig. 2.10). The frequency at which

$$|K_r(j\omega)K_o(j\omega)| = 1 \tag{2.57}$$

is called the gain crossover frequency.

The closed-loop system behaviour is considered to be good, in the sense of having good gain and phase margins, if simultaneously crossover frequencies are as high as possible and

- the gain margin is not less than 6 to 8 dB;
- the phase margin is not less than $\frac{\pi}{3}$ [rad].

The classical gain and phase margins are less conservative measures of the plant stability margin than the additive or multiplicative robust stability margins.

2.4.2 Sensitivity and complementary sensitivity

An example of the sensitivity function is presented in Fig. 2.11. From engineering experience of applying the sensitivity function in control system design, some heuristic recommendations follow. Desirable control properties are achieved if for low frequencies the graph of $|S(j\omega)|$ is flat and its values are close to zero. This ensures high steady-state accuracy and small distortion of the reference signal over a broad frequency band. When high-frequency disturbances are negligible, an increase of the resonance frequency ω_{rq} is

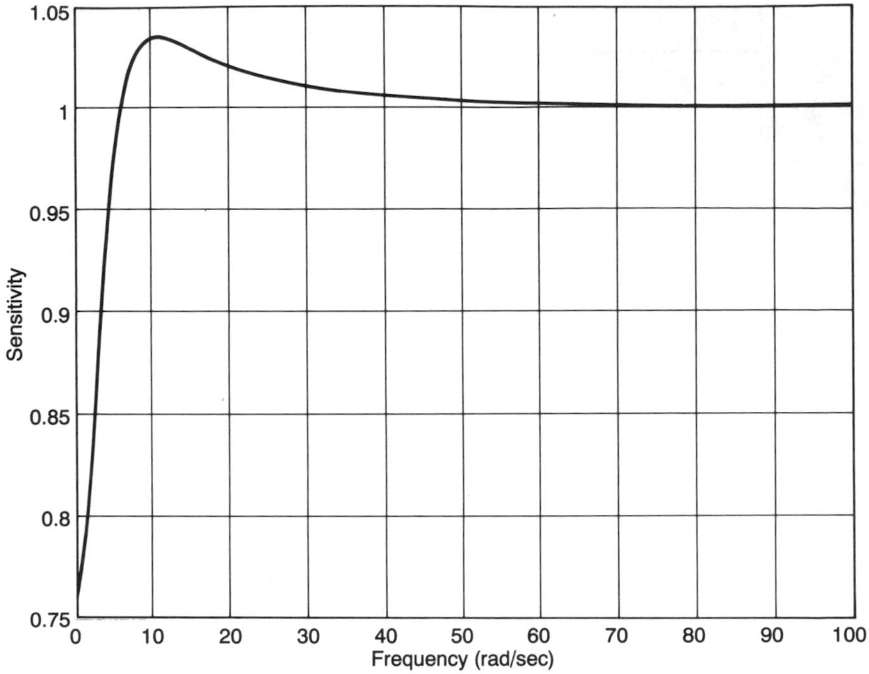

Figure 2.11. Sensitivity function

profitable. The bandwidth of the disturbance stop band is enlarged, equivalent to increase of the bandwidth of the reference signal pass band. The settling time becomes small. The proper value of the stability margin should be preserved.

The complementary sensitivity function is the gain of the closed-loop transfer function (2.12). The following two methods of controller design using $|M(j\omega)|$ can be found in engineering practice:

- the optimal gain criterion;

- the M_{max} criterion.

The specifications of above methods are heuristic.

In the optimal gain method the values of the $|M(j\omega)|$ characteristic should roll off (possibly slowly) with the increase of the frequency ω without any resonance peak. The pass band should be as large as required. The time response of the control system designed with such recommendations is characterized by short settling time and small overshoot.

The closed-loop system designed using the M_{max} criterion is considered to be good if, simultaneously, the value of the resonance peak M_{max} is in the range $[1.2, 1.3]$ and the upper bound of the reference signal pass band is as high as possible.

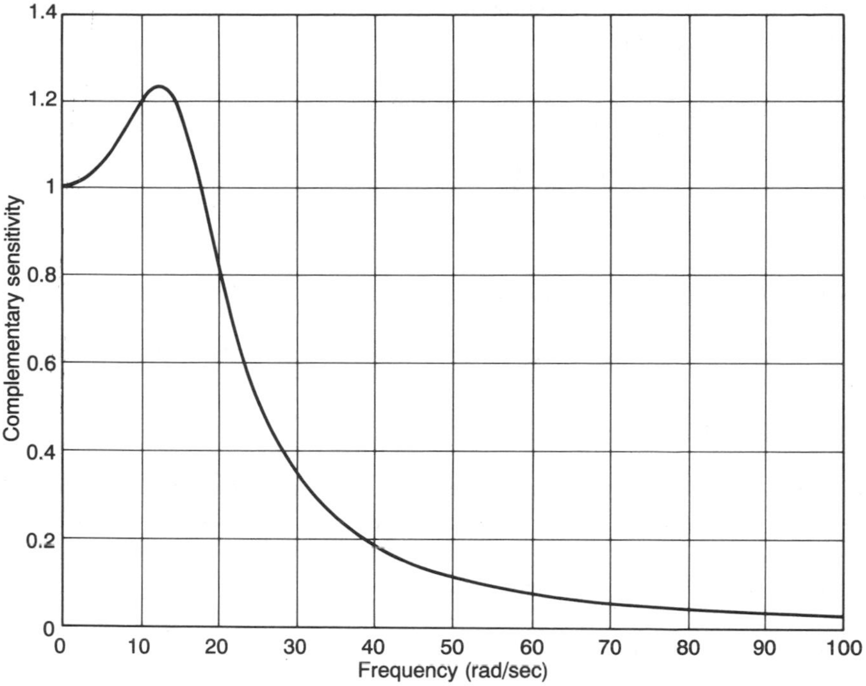

Figure 2.12. Complementary sensitivity function

2.4.3 Compensation

Application of a compensator gives the ability to tune dynamical properties by changing the crossover gain and frequency, as well as the bandwidth of the closed-loop system. It can also be used for the cancellation of stable pole-zero pairs of systems for which closed-loop characteristics are not sensitive to variations of pole and zero locations. The following three compensators are commonly used:

- PI or lag compensator, with the transfer function:

$$K_k(s) = \frac{1 + sT}{1 + s\mu T}. \tag{2.58}$$

where $\mu > 1$. This compensator increases the gain margin about μ times with a small decrease of the crossover frequency.

- PD or lead compensator, with the transfer function:

$$K_k(s) = \frac{1 + sT}{1 + s\dfrac{T}{\mu}}. \tag{2.59}$$

where $\mu > 1$. This increases the gain crossover frequency. There is a reduction of the settling time. The value of μ should be chosen very carefully because large values of μ could result in amplification of disturbances. This compensator is not used for plants with large time delays.

- PID or lead-lag compensator, with the transfer function:

$$K_k(s) = \frac{(1+sT_1)(1+sT_2)}{(1+s\mu T_1)(1+s\frac{T_2}{\mu})}. \quad (2.60)$$

  where $\mu > 1$. This compensator combines features of the lead and lag compensators.

The value of μ in the compensator can be tuned using the gain and phase margins criteria.

2.4.4 PID control

Dynamical system responses and steady-state accuracy are more effectively determined by controllers. Years of experience have established an old-fashioned proportional-integral-derivative control (PID) as the technique most often used in engineering practice. The following structures are commonly used:

- P or proportional controller with the transfer function:

$$K_r(s) = K_p. \quad (2.61)$$

  The proportional controller has a rather modest ability for shaping the sensitivity and complementary sensitivity functions. The increase of K_p improves the steady-state accuracy and low-frequency disturbance rejection, reduces the stability margin and magnifies any resonance peaks in $|S(j\omega)|$.

- PI or proportional-integral controller with the transfer function:

$$K_r(s) = K_p\left(1+\frac{1}{sT_I}\right). \quad (2.62)$$

  The proportional-integral controller simultaneously improves the low-frequency disturbance rejection and reduces the steady-state error, the disturbance rejection bandwidth and resonance frequency. The value T_I should be chosen such that at resonance frequency with P controller the integral action introduces only small phase lag.

- PD or proportional-derivative controller with the transfer function:

$$K_r(s) = K_p(1+sT_D). \quad (2.63)$$

  The proportional-derivative controller enlarges the disturbance rejection bandwidth, does not reduce the steady-state error and amplifies high-frequency disturbances.

- PID or proportional-integral-derivative controller with the transfer function:

$$K_r(s) = K_p \left(1 + \frac{1}{sT_I} + sT_D\right). \tag{2.64}$$

The proportional-integral-derivative controller combines features of the PI and PD controllers. The values of T_I and T_D should be chosen so that at the resonance frequency with the P controller the integral-derivative action introduces phase lag in the range $[\frac{\pi}{6}, \frac{\pi}{4}]$.

For plants which can be approximated by a first-order dynamical system with time delay:

$$K_o(s) = e^{-sT_0}\frac{1}{1+sT}, \tag{2.65}$$

the Ziegler–Nichols method for tuning PID controllers is recommended. In this case the closed-loop system step response is characterized by a damping ratio close to 0.5. This method is based on a stability analysis. The tuning of a PID controller is possible without knowledge of the plant model. For a closed-loop system with a proportional controller, the gain K_P is increased until the system becomes oscillatory. The oscillation frequency is denoted by ω_m and the corresponding gain of the P controller by K_m. Using this information the parameters of the PID controllers are chosen as in Table 2.1.

Table 2.1. The parameters of Ziegler–Nichols controllers

Controller	K_p	T_I	T_D
P	$0.5K_m$	–	–
PI	$0.45K_m$	$\frac{10\pi^2}{3\omega_m^2}$	–
PID	$0.6K_m$	$\frac{\pi}{\omega_m}$	$\frac{\pi}{4\omega_m}$

Determination of ω_m and K_m can make use of the root locus and the Nyquist plot of the open-loop transfer function. The frequency of oscillations and corresponding gain can be read off from the point at which root locus crosses the $j\omega$ axis, or can be evaluated from the gain margin and phase crossover frequency.

PID controllers can be also tuned for a given response using optimization methods with different cost functions. The most popular cost functions are as follows:

- Integral of the Absolute value of Error (IAE):

$$IAE = \int_0^\infty |e(t)|\, dt; \tag{2.66}$$

- Integral of Time multiplied by the Absolute value of Error (ITAE):

$$ITAE = \int_0^\infty t\,|e(t)|\, dt; \tag{2.67}$$

- Integral of the Square Error (ISE):

$$ISE = \int_0^{\infty} e^2(t)dt. \tag{2.68}$$

For a first-order dynamical system with time delay and step changes in the reference signal or disturbances, the parameters of the PID controller that minimize the above criteria are tabulated in various publications starting from the 1950s.

2.4.5 Pole placement

The output signal of a tracking system should follow a reference signal with minimum error in the transient and steady state. The transient response of a system is determined by its closed-loop pole locations. To achieve any desired pole locations a state feedback can be used .

Consider a linear system represented by the nth-order state-space equations:

$$\dot{\boldsymbol{x}}(t) = \boldsymbol{A}\boldsymbol{x}(t) + \boldsymbol{b}u(t) \tag{2.69}$$

$$y(t) = \boldsymbol{c}\boldsymbol{x}(t) + du(t). \tag{2.70}$$

which is controllable and observable. The dynamical behaviour of the closed-loop system is modified by state feedback:

$$u(t) = \boldsymbol{k}^T \boldsymbol{x}(t). \tag{2.71}$$

Variations of the feedback gain vector $\boldsymbol{k}^T$ result in changes of location of the characteristic equation roots:

$$\det\left(s\boldsymbol{I} - \boldsymbol{A} + \boldsymbol{b}\boldsymbol{k}^T\right) = 0. \tag{2.72}$$

From the desired closed-loop pole locations follow coefficients β_i $(i = 0, 1, \ldots, n-1)$ of the corresponding characteristic polynomial:

$$W(s) = s^n + \beta_{n-1}s^{n-1} + \cdots + \beta_1 s + \beta_0. \tag{2.73}$$

The unique solution of the equation

$$\det\left(s\boldsymbol{I} - \boldsymbol{A} + \boldsymbol{b}\boldsymbol{k}^T\right) = W(s) \tag{2.74}$$

is of the following feedback gain law form:

$$\boldsymbol{k}^T = [0, 0, \ldots, 0, 1]\boldsymbol{\Theta}^{-1}\boldsymbol{W}(\boldsymbol{A}), \tag{2.75}$$

where:

$$\boldsymbol{W}(\boldsymbol{A}) = \boldsymbol{A}^n + \beta_{n-1}\boldsymbol{A}^{n-1} + \cdots + \beta_1\boldsymbol{A} + \beta_0\boldsymbol{1}, \tag{2.76}$$

$$\boldsymbol{\Theta} = \left[\boldsymbol{b}, \boldsymbol{A}\boldsymbol{b}, \boldsymbol{A}^2\boldsymbol{b}, \ldots, \boldsymbol{A}^{n-1}\boldsymbol{b}\right]. \tag{2.77}$$

Note that the state feedback is able to shift poles only. The location of the zeros is not changed by this kind of feedback.

In the case of multi-input systems there is no unique solution to the pole-placement problem. Equation (2.74) has additional degrees of freedom which can be used for minimization of the sensitivity of the closed-loop pole perturbations.

2.5 MULTIVARIABLE PLANT

2.5.1 Representation and uncertainty

Many real plants have more than one input and one output signal. Signals are gathered into vectors and systems are called multivariable. Mathematically, relations between output vector $\boldsymbol{y}(t)$ (dim $\boldsymbol{y}(t) = n$) and input vector $\boldsymbol{u}(t)$ (dim $\boldsymbol{u}(t) = n$) are described using the Laplace transform by a transfer function matrix $\boldsymbol{K}_o(s)$:

$$\boldsymbol{Y}(s) = \boldsymbol{K}_o(s)\boldsymbol{U}(s), \tag{2.78}$$

where:

$$\boldsymbol{K}_o(s) = \begin{bmatrix} K_{11}(s) & K_{12}(s) & \dots & K_{1n}(s) \\ K_{21}(s) & K_{22}(s) & \dots & K_{2n}(s) \\ \vdots & \vdots & \vdots & \vdots \\ K_{m1}(s) & K_{m2}(s) & \dots & K_{nn}(s) \end{bmatrix}. \tag{2.79}$$

It is assumed that the number of controlled variables (outputs of the plant) is equal to the number of manipulated variables (inputs of the plant).

Each element $K_{ij}(s)$ of $\boldsymbol{K}_o(s)$ is an ordinary transfer function between the output signal j and input signal i. The graphical representation of the transfer-function matrix in the frequency domain is an array of components, each a frequency response graph. This representation of $\boldsymbol{K}_o(j\omega)$ is called the Nyquist array.

The state-space plant representation:

$$\dot{\boldsymbol{x}}(t) = \boldsymbol{A}\boldsymbol{x}(t) + \boldsymbol{B}\boldsymbol{u}(t) \tag{2.80}$$

$$\boldsymbol{y}(t) = \boldsymbol{C}\boldsymbol{x}(t) + \boldsymbol{D}\boldsymbol{x}(t). \tag{2.81}$$

is connected to the transfer-function matrix by the following relation:

$$\boldsymbol{K}_o(s) = \boldsymbol{C}\,(s\boldsymbol{I} - \boldsymbol{A})^{-1}\,\boldsymbol{B} + \boldsymbol{D}. \tag{2.82}$$

Uncertainty of multivariable plant is modelled in the same way as for the SISO case. Structured and unstructured uncertainty models are used. The unstructured uncertainty is also categorized as being either additive or multiplicative. The multiplicative uncertainty is able to be located at the system input or output (Fig. 2.13). In sequence the following unstructured uncertainty models are used:

- the additive uncertainty $\boldsymbol{\Delta}_a(s)$:

$$\tilde{\boldsymbol{K}}_o(s) = \boldsymbol{K}_o(s) + \boldsymbol{\Delta}_a(s); \tag{2.83}$$

- the input multiplicative uncertainty $\boldsymbol{\Delta}_{mi}(s)$:

$$\tilde{\boldsymbol{K}}_o(s) = \boldsymbol{K}_o(s)\left(\boldsymbol{I} + \boldsymbol{\Delta}_{mi}(s)\right); \tag{2.84}$$

- the output multiplicative uncertainty $\boldsymbol{\Delta}_{mo}(s)$:

$$\tilde{\boldsymbol{K}}_o(s) = \left(\boldsymbol{I} + \boldsymbol{\Delta}_{mi}(s)\right)\boldsymbol{K}_o(s). \tag{2.85}$$

In these equations $\tilde{\boldsymbol{K}}_o(s)$ denotes the actual model of the plant.

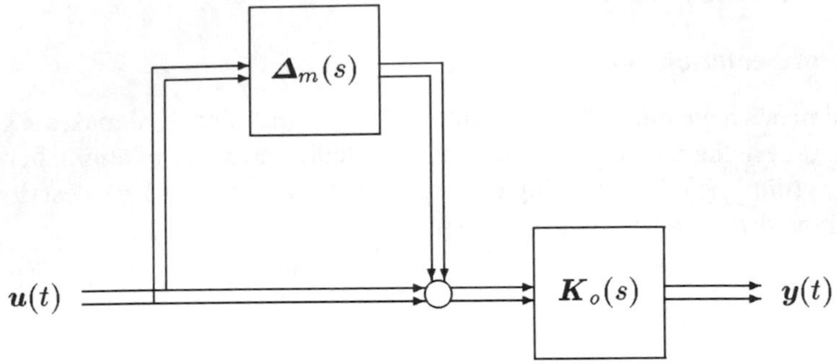

(a)

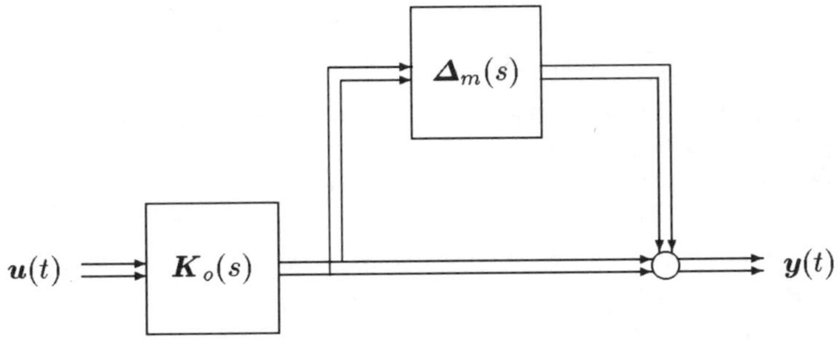

(b)

Figure 2.13. Multivariable multiplicative uncertainty at the system input (a) and output (b)

2.5.2 Interaction

The structure of the transfer-function matrix determines the plant interaction between variables. As a result of an interaction, a change in one input signal acts on more than one output signal. In terms of the transfer-function matrix, interaction is represented by the off-diagonal elements. If the transform function matrix is diagonal, then the corresponding plant does not exhibit interaction. A suitable measure of interaction in the frequency domain are the Gershgorin bands on the diagonal elements of the Nyquist array of $\boldsymbol{K}_o(j\omega)$. For each frequency ω and a diagonal element i ($i = 1, 2, \ldots, n$) of the transfer-function matrix $\boldsymbol{K}_o(j\omega)$, the radius:

$$r(\omega) = \sum_{l=1, l\neq i}^{n} |K_{il}(j\omega)| \tag{2.86}$$

is calculated and a circle of this radius with centre at $K_{ii}(j\omega)$ is plotted. A whole set of such circles for each diagonal frequency response forms the Gershgorin bounds. The width of those bounds gives a visual measure of interaction: wide bounds indicate lots of interaction.

If for all frequencies $0 \leq \omega < \infty$ the condition

$$|K_{ii}(j\omega)| > \sum_{l=1, l\neq i}^{n} |K_{il}(j\omega)| \tag{2.87}$$

is satisfied, the transfer-function matrix $\boldsymbol{K}_o(j\omega)$ is diagonally column dominant. The row dominance is defined in the same way. Diagonal dominance of the transfer-function matrix can be checked graphically by examining the Gershgorin bounds. If each of these bounds for all diagonal elements of $\boldsymbol{K}_o(j\omega)$ excludes the origin, then $\boldsymbol{K}_o(j\omega)$ is diagonally dominant. In the most cases it guarantees a low level of interaction.

The interaction can be reduced by decoupling techniques. Theoretically, they remove control loop interaction. The reduction of control loop interaction is achieved by an additional controller called the decoupler. The decoupler is added to the plant. In engineering practice, the benefits of the decoupling technique depend on the accuracy of the plant model and are not fully realized.

2.5.3 Signal and system measures

The components of input and output vectors are the time-domain functions. To measure their size, the following definitions of norms are used:

- the L_2-norm:

$$||\boldsymbol{u}(t)||_2 = \sqrt{\int_0^\infty \boldsymbol{u}^T(t)\boldsymbol{u}(t)dt}; \tag{2.88}$$

- the L_∞-norm:

$$||\boldsymbol{u}(t)||_\infty = \sup_{t\geq 0} \max_{r=1,2,\ldots,n} |u_r(t)|, \tag{2.89}$$

where $\boldsymbol{u}(t) = \left[u_1(t), u_2(t), \ldots, u_n(t)\right]$.

The performance of SISO systems with feedback is determined by the variation of the gain with frequency; the disturbance rejection and accuracy of the tracking depend only on the open-loop gain. In the multivariable case the concept of the gain is replaced by singular values of the transfer-function matrix, which are called the principal gains.

The singular values $\sigma_1(\omega), \sigma_2(\omega), \ldots, \sigma_n(\omega)$ of $\boldsymbol{K}_o(s)$ are functions of the frequency ω. Their plots are generalizations of Bode magnitude plots for multivariable systems. The largest singular value ($\overline{\sigma}(\omega)$) and the smallest singular value ($\underline{\sigma}(\omega)$) are measures of changes of the plant amplification and attenuation with frequency ω, respectively.

The largest singular value of the matrix $\boldsymbol{K}_o(s)$ is its spectral norm. This changes with the frequency ω. The transfer-function matrix can also be characterized by a single non-negative number using the following system norms:

- the H_2-norm:

$$||\boldsymbol{K}_o(j\omega)||_2 = \sqrt{\frac{1}{2\pi}\int_{-\infty}^{\infty} tr\left\{\boldsymbol{K}_o(j\omega)\boldsymbol{K}_o^T(-j\omega)\right\} d\omega}$$

$$= \sqrt{\frac{1}{2\pi}\int_{-\infty}^{\infty}\sum_{r=1}^{n}\sigma_r^2(\omega)d\omega}, \tag{2.90}$$

where $tr\{\cdot\}$ denotes the trace of the matrix;

- the H_∞-norm:

$$||\boldsymbol{K}_o(j\omega)||_\infty = \sup_\omega \overline{\sigma}(\omega). \tag{2.91}$$

The value of $||\boldsymbol{K}_o(j\omega)||_\infty$ can be read off from a plot of the largest principal gain $\overline{\sigma}(\boldsymbol{K}_o(j\omega))$.

The sub-multiplicative property is fulfilled only by the H_∞-norm. This property is not satisfied by the H_2-norm.

The ratio $(\frac{\overline{\sigma}(0)}{\underline{\sigma}(0)})$ of the largest singular value of the transfer-function matrix to the smallest one is a condition number of the gain matrix. It is a measure of the difficulty of the multivariable control problem. Large condition numbers usually indicate that the degrees of freedom of the system are such that all requirements relating to the control system being designed will be difficult or impossible to meet.

2.6 MULTIVARIABLE FEEDBACK SYSTEMS

2.6.1 Performance

Multivariable plants are controlled by multi-input and multi-output controllers which are inserted into the negative multivariable feedback loop. There are two basic structures of multivariable controllers:

- the cross-coupled structure for which the controller transfer-function matrix $\boldsymbol{K}_r(s)$ off-diagonal elements are not all equal to 0;
- the decentralized structure for which the controller transfer-function matrix $\boldsymbol{K}_r(s)$ is diagonal.

The multivariable closed-loop system block diagram is shown in Fig. 2.14, where:

$\boldsymbol{K}_o(s)$ – the plant transfer-function matrix;

$\boldsymbol{K}_r(s)$ – the controller transfer-function matrix;

$\boldsymbol{y}(t)$ – the vector of output signals;

$\boldsymbol{y}_z(t)$ – the vector of reference signals;

$\boldsymbol{d}(t)$ – the vector of disturbance signal at the output;

$\boldsymbol{n}(t)$ – the vector of measurement noise;

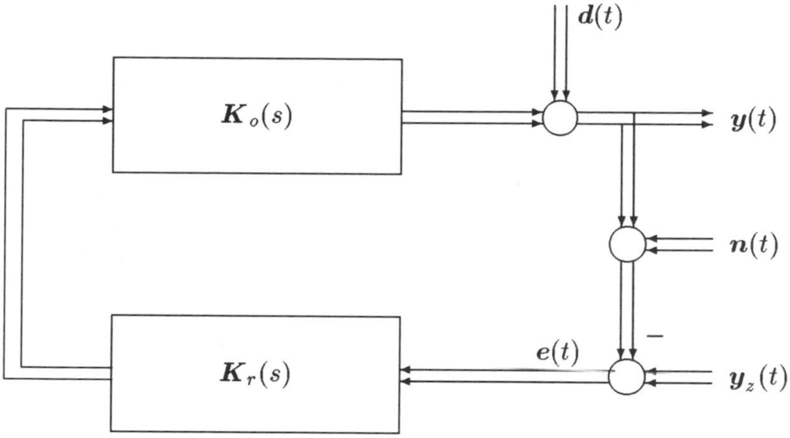

Figure 2.14. Multivariable closed-loop system

$e(t)$ – the vector of tracking errors.

Using the Laplace transforms, the output vector $y(t)$ of the closed-loop system can be represented as:

$$\begin{aligned} \boldsymbol{Y}(s) &= (\boldsymbol{I} + \boldsymbol{K}_o(s)\boldsymbol{K}_r(s))^{-1}\boldsymbol{K}_o(s)\boldsymbol{K}_r(s)\left[\boldsymbol{Y}_z(s) - \boldsymbol{N}(s)\right] \\ &\quad + (\boldsymbol{I} + \boldsymbol{K}_o(s)\boldsymbol{K}_r(s))^{-1}\boldsymbol{D}(s). \end{aligned} \tag{2.92}$$

The first factor:

$$\boldsymbol{M}(s) = (\boldsymbol{I} + \boldsymbol{K}_o(s)\boldsymbol{K}_r(s))^{-1}\boldsymbol{K}_o(s)\boldsymbol{K}_r(s) \tag{2.93}$$

is the closed-loop transfer function, known as the complementary sensitivity function. The second one:

$$\boldsymbol{S}(s) = (\boldsymbol{I} + \boldsymbol{K}_o(s)\boldsymbol{K}_r(s))^{-1} \tag{2.94}$$

is called the sensitivity function. The closed-loop requirements can be expressed in terms of their principal gains:

- to reduce the influence of disturbances on the output signal, the sensitivity function measured as $\overline{\sigma}\left(\boldsymbol{S}(j\omega)\right)$ should be as small as possible for frequencies present in disturbances;

- to reduce the influence of the measurement noise on the output signal, the complementary sensitivity function measured as $\overline{\sigma}\left(\boldsymbol{M}(j\omega)\right)$ should be as small as possible for frequencies present in the measurement noise;

- to reduce the tracking error, the complementary sensitivity function measures $\underline{\sigma}\left(\boldsymbol{M}(j\omega)\right)$ and $\overline{\sigma}\left(\boldsymbol{M}(j\omega)\right)$ should be close to 1 for frequencies present in the reference signal.

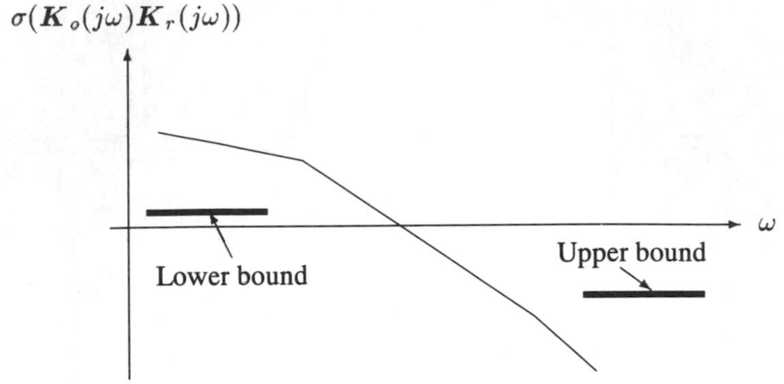

Figure 2.15. The open-loop gain specification

These requirements can be transformed into conditions which should be satisfied by the open-loop transfer-function matrix principal gains. The desirable shape of the open-loop transfer-function gain measured as $\overline{\sigma}(\boldsymbol{K}_o(j\omega)\boldsymbol{K}_r(j\omega))$ and $\underline{\sigma}(\boldsymbol{K}_o(j\omega)\boldsymbol{K}_r(j\omega))$ is represented in Fig. 2.15. A low level of the sensitivity function and small tracking error are ensured if $\underline{\sigma}(\boldsymbol{K}_o(s)\boldsymbol{K}_r(s))$ and $\underline{\sigma}(\boldsymbol{K}_o(j\omega)\boldsymbol{K}_r(j\omega))$ are large. This gives the lower bound for the open-loop transfer-function matrix gain at low frequencies and conflicts with the minimization of a control energy $\overline{\sigma}(\boldsymbol{K}_r(s))$ and the transmission of measurement noise. It implies that at high frequencies the open-loop gain should be located below an upper bound.

These control system objectives can be reformulated using frequency-dependent bounds on the sensitivity and complementary sensitivity functions, and norms. Bounds are approximated by gains of transfer functions $W_S(s)$ and $W_M(s)$ that are chosen in an arbitrary manner. The sensitivity and complementary sensitivity functions performance are represented as:

$$\overline{\sigma}\left(W_S^{-1}(j\omega)\boldsymbol{S}(j\omega)\right) \leq 1, \tag{2.95}$$

$$\overline{\sigma}\left(W_M^{-1}(j\omega)\boldsymbol{M}(j\omega)\right) \leq 1. \tag{2.96}$$

It is worth noting that transfer functions $W_S(s)$ and $W_M(s)$ are scalar. In general, for the H_∞-norm problem representation these frequency-dependent bounds are defined by matrices $\boldsymbol{W}_S(j\omega)$ and $\boldsymbol{W}_M(j\omega)$:

$$||\boldsymbol{W}_S(j\omega)\boldsymbol{S}(j\omega)||_\infty \leq 1, \tag{2.97}$$

$$||\boldsymbol{W}_M(j\omega)\boldsymbol{M}(j\omega)||_\infty \leq 1. \tag{2.98}$$

2.6.2 Stability

The stability of a multivariable closed-loop system can be analysed in the same way as for SISO systems, i.e. by checking the locations of the characteristic equation roots. If all roots lie in the left-half s-plane, the corresponding system is stable.

The characteristic equation of a multivariable system can be defined in terms of the state-space description as:

$$\det(s\boldsymbol{I} - \boldsymbol{A}) = 0; \tag{2.99}$$

or using the transfer-function matrices by:

$$\det\left(\boldsymbol{I} + \boldsymbol{K}_o(s)\boldsymbol{K}_r(s)\right)\delta(s) = 0, \tag{2.100}$$

where $\delta(s)$ is a product of all denominators of the transfer-function matrix $\boldsymbol{K}_o(s)\boldsymbol{K}_r(s)$.

The classical Routh or Hurwitz tests are easy to perform when the characteristic equation is of the form of the polynomial equation (2.35). Closed-loop systems with time delays can be handled using generalizations of the Nyquist stability theorem.

It is easy to show that:

$$\det\left(\boldsymbol{I} + \boldsymbol{K}_o(j\omega)\boldsymbol{K}_r(j\omega)\right) = \prod_i^n \left(1 + \lambda_i(j\omega)\right), \tag{2.101}$$

where $\lambda_i(j\omega)$ are frequency-dependent eigenvalues of the frequency response matrix $\boldsymbol{K}_o(j\omega)\boldsymbol{K}_r(j\omega)$. A set of graphs of all $\lambda_i(j\omega)$ in polar coordinates for frequencies varying from $-\infty$ to ∞ is called the characteristic loci. It enables the stability analysis with the generalized multivariable Nyquist theorem which states that the closed-loop system is asymptotically stable if and only if the characteristic loci of $\boldsymbol{K}_o(j\omega)\boldsymbol{K}_r(j\omega)$ for $-\infty < \omega < \infty$ encircle the $-1 + j0$ point as many times anti-clockwise as $\boldsymbol{K}_o(j\omega)\boldsymbol{K}_r(j\omega)$ has right-half s-plane poles.

The number of encirclements of the $-1 + j0$ point made by the characteristic loci can also be evaluated by using the Nyquist array with the Gershgorin bounds by counting the encirclements of the $-1 + j0$ point by the Gershgorin bounds, so long as this point is not within any of the Gershgorin bounds.

The internal stability of the closed-loop system is defined in the same way as for the SISO case. Internal signals $\boldsymbol{e}(t)$, $\boldsymbol{u}(t)$ and $\boldsymbol{y}(t)$ will remain bounded for all bounded input signals $\boldsymbol{y}_z(t)$, $\boldsymbol{d}(t)$ and $\boldsymbol{n}(t)$ if all transfer-functions matrices between inputs and outputs are stable. The multivariable feedback system satisfying this requirement is said to be internally stable.

2.6.3 Robust stability

Conditions for robust stability of a multivariable control system can be derived from a multivariable version of the small gain theorem. It states that the closed-loop system will remain stable if a gain measure of the product of all transfer-function matrices constituting the feedback is less than 1.

For a stable plant and uncertainty transfer-function matrices $\boldsymbol{\Delta}_a(s)$, $\boldsymbol{\Delta}_{mi}(s)$ and $\boldsymbol{\Delta}_{mo}(s)$, the closed-loop system will be robustly stable if for all frequencies ω:

- the additive uncertainty satisfies:

$$\overline{\sigma}\left(\boldsymbol{S}(j\omega)\boldsymbol{K}_r(j\omega)\boldsymbol{\Delta}_a(j\omega)\right) < 1; \tag{2.102}$$

- the input multiplicative uncertainty satisfies:

$$\overline{\sigma}\left(\left(\boldsymbol{I}+\boldsymbol{K}_r(j\omega)\boldsymbol{K}_o(j\omega)\right)^{-1}\boldsymbol{K}_r(j\omega)\boldsymbol{K}_o(j\omega)\boldsymbol{\Delta}_{mi}(j\omega)\right) < 1; \qquad (2.103)$$

- the output multiplicative uncertainty satisfies:

$$\overline{\sigma}\left(\boldsymbol{M}(j\omega)\boldsymbol{\Delta}_{mo}(j\omega)\right) < 1. \qquad (2.104)$$

The size of the smallest uncertainty that destabilizes the closed-loop system is

- the additive robust stability margin:

$$\boldsymbol{ASM} = \frac{1}{||(\boldsymbol{S}(j\omega)\boldsymbol{K}_r(j\omega)||_\infty}; \qquad (2.105)$$

- the input multiplicative robust stability margin:

$$\boldsymbol{MSM}_i = \frac{1}{||\left(\boldsymbol{I}+\boldsymbol{K}_r(j\omega)\boldsymbol{K}_o(j\omega)\right)^{-1}\boldsymbol{K}_r(j\omega)\boldsymbol{K}_o(j\omega)||_\infty}; \qquad (2.106)$$

- the output multiplicative robust stability margin:

$$\boldsymbol{MSM}_o = \frac{1}{||\boldsymbol{M}(j\omega)||_\infty}. \qquad (2.107)$$

In the case of a multivariable plant there is a difference between the input and output multiplicative uncertainties. There is a possibility of obtaining good robust performance for the input (output) uncertainty model but poor robustness at the output (input). In the SISO case such a difference is not distinguishable because transfer functions of the plant with input or output multiplicative uncertainty are equal.

2.7 MULTIVARIABLE CONTROL DESIGN

2.7.1 Control structure design

For convenience, we assume that a decentralized controller is under consideration. The plant interaction between inputs and outputs makes the selection of the best pairing of inputs and outputs for multivariable plants a difficult task. This analysis, undertaken at an early stage of the control system design, may avoid many problems of input and output specification which satisfy quality requirements. An incorrect pairing very often results in a poor closed-loop system behaviour. The main information enabling appropriate pairing should come from a physical description of the plant. The theory of control systems gives two useful measures of the plant interaction based only on the steady-state gain matrix $\boldsymbol{K}_o(0)$:

- the relative gain array ($\boldsymbol{RGA}$);

$$\boldsymbol{RGA} = \boldsymbol{K}_o(0).*\left(\boldsymbol{K}_o^{-1}(0)\right)^T, \tag{2.108}$$

where $*$ denotes the Hadamard product;

- the Niederliński index (Ni):

$$Ni = \frac{\det \boldsymbol{K}_o(0)}{\prod_{i=0}^{n} K_{ii}(0)}. \tag{2.109}$$

The relative gain array is invariant to input and output scaling and each row or column sum of its elements is equal to 1. Permutations of rows or columns of the transfer-function matrix result in the same permutations of rows or columns of $\boldsymbol{RGA}$. A design recommendation of input–output pairing is that the corresponding diagonal elements of the relative gain array are positive and as close to 1 as possible. Large or negative elements of $\boldsymbol{RGA}$ result in difficulties in controlling the plant. The disadvantage of the relative gain array approach is that it ignores plant dynamics. Pairings proposed by the relative gain array should be additionally tested for stability using the Niederliński index.

Assuming that each feedback controller contains an integral action, each individual control loop is stable when any of other loops are opened, and that all elements of the transfer-function matrix are stable, a sufficient condition for instability is that the Niederliński index is less than zero. If this condition is not satisfied, the closed-loop system may or may not be unstable depending on numerical values of the controller settings. For two-input and two-output systems this test is the necessary and sufficient condition.

2.7.2 Diagonal control

Over 90% of industrial controllers are of PID type. Their tuning in the multivariable case utilizes the well established Ziegler–Nichols settings and the generalization of the stability margin for characteristic loci.

The simplest approach is to ignore the multivariable nature of the plant. A SISO controller is designed for one input–output pair and the corresponding loop is closing. This algorithm is repeated for the remaining input-output pairs, ensuring that the effects of the loop being closed is taken into account and the closed-loop system is stable at each design stage. This method is known as the sequential loop closing.

PID diagonal controllers can also be tuned by using results of experiments which are generalizations of the Ziegler–Nichols experiment with all loops closed. The sample tuning procedure for PID controllers is as follows:

1. Assume values of n coefficients c_i which correspond to given relative control quality requirements for separate output variables.

2. Close the feedback loop with the diagonal controller, leaving as its components proportional SISO controllers. Find the frequency Ω_m of oscillations and the corresponding set of $K_{P,i}$ coefficients of the P-type controllers satisfying the constraints:

$$\frac{K_{P,i}K_{ii}(0)}{K_{P,i+1}K_{(i+1)(i+1)}(0)} = \frac{c_i}{c_{i+1}}, \tag{2.110}$$

where $i = 1, 2, \ldots, n$.

3. Calculate the Ziegler–Nichols PID controller settings using Ω_m and the set of $K_{P,i}$. Change controllers K_P by a factor a $(0.2 < a < 0.6)$. The choice of a depends on the relation of Ω_m to the oscillation frequencies for all subsequent plants. In the case of large differences between those frequencies the upper bound of an a is recommended. When those values are comparable, a value of a close to the lower bound should be used.

4. If relative control quality requirements are not satisfied, then return to step 1 and change the set of coefficients c_i.

Determination of Ω_m and the values of $K_{P,i}$ $(i = 1, 2, \ldots, n)$ is possible on the basis of the gain margin and corresponding crossover frequency calculated from the characteristic locus graph.

Quality requirements are very often expressed in terms of the sensitivity or complementary sensitivity functions or the gain and phase margins calculated from the characteristic loci.

2.7.3 The characteristic locus method

The basic concepts of this method come from SISO frequency response shaping. The spectral decomposition of the plant transfer-function matrix is of the form:

$$\boldsymbol{K}_o(j\omega) = \boldsymbol{T}(j\omega)\boldsymbol{\Lambda}(j\omega)\boldsymbol{T}^{-1}(j\omega), \tag{2.111}$$

where $\boldsymbol{T}(j\omega)$ is a matrix of eigenvectors and $\boldsymbol{\Lambda}(j\omega)$ is the diagonal matrix of the plant eigenvalues:

$$\boldsymbol{\Lambda}(j\omega) = diag\left\{\lambda_1(j\omega), \lambda_2(j\omega), \ldots, \lambda_n(j\omega)\right\}. \tag{2.112}$$

It is assumed that a controller is of the structure:

$$\boldsymbol{K}_r(j\omega) = \boldsymbol{T}(j\omega)\boldsymbol{\Xi}(j\omega)\boldsymbol{T}^{-1}(j\omega), \tag{2.113}$$

where $\boldsymbol{\Xi}(j\omega)$ is the diagonal matrix of controller eigenvalues:

$$\boldsymbol{\Xi}(j\omega) = diag\left\{\xi_1(j\omega), \xi_2(j\omega), \ldots, \xi_n(j\omega)\right\}. \tag{2.114}$$

The open-loop transfer-function matrix is then:

$$\boldsymbol{K}_o(j\omega)\boldsymbol{K}_r(j\omega) = \boldsymbol{T}(j\omega)\boldsymbol{\Lambda}(j\omega)\boldsymbol{\Xi}(j\omega)\boldsymbol{T}^{-1}(j\omega), \tag{2.115}$$

where:

$$\boldsymbol{\Lambda}(j\omega)\boldsymbol{\Xi}(j\omega) = diag\left\{\Lambda_1(j\omega)\xi_1(j\omega), \Lambda_2(j\omega)\xi_2(j\omega), \ldots, \Lambda_n(j\omega)\xi_n(j\omega)\right\}. \tag{2.116}$$

It follows that the eigenvalues of $\boldsymbol{K}_o(j\omega)\boldsymbol{K}_r(j\omega)$ are products of the plant and controller eigenvalues, which form the fundamentals of the characteristic locus methods. For each plant eigenvalue $\lambda_i(j\omega)$ the corresponding controller eigenvalue $\xi_i(j\omega)$ can be designed separately using classical methods and the graph of the characteristic loci.

Very often matrices $\boldsymbol{T}(j\omega)$ and $\boldsymbol{T}^{-1}(j\omega)$ have no real-world counterparts and must be replaced by realizable approximations:

$$\begin{aligned} \boldsymbol{A}(j\omega) &\approx \boldsymbol{T}(j\omega), & (2.117)\\ \boldsymbol{B}(j\omega) &\approx \boldsymbol{T}^{-1}(j\omega). & (2.118) \end{aligned}$$

The best case is when $\boldsymbol{A}(j\omega)$ and $\boldsymbol{B}(j\omega)$ are independent, real-valued matrices because they can easily be implemented by nets of amplifiers.

2.7.4 The Nyquist array method

The direct Nyquist array method reduces loop interaction by determining a controller $\boldsymbol{K}_r(s)$ such that the transfer-function matrix $\boldsymbol{K}_o(j\omega)\boldsymbol{K}_r(j\omega)$ is diagonally dominant. The diagonal dominance can be achieved by using elementary row or column operations. They result in the following structure of the transfer-function matrix $\boldsymbol{K}_r(j\omega)$:

$$\boldsymbol{K}_r(s) = \boldsymbol{K}_a\boldsymbol{K}_b(s)\boldsymbol{K}_c(s), \quad (2.119)$$

where:

- $\boldsymbol{K}_a$ is a permutation matrix, which reorders the inputs by interchanging columns of $\boldsymbol{K}_o(s)$;
- $\boldsymbol{K}_b(s)$ is a product of elementary matrices, which adds one column to another with scaling factor;
- $\boldsymbol{K}_c(s)$ is the diagonal matrix, which represents a set of single-loop controllers.

In engineering practice, it is worthwhile achieving diagonal dominance with the matrix $\boldsymbol{K}_a\boldsymbol{K}_b(s)$ independent of s. Such controllers are easy to implement. A simple method of achieving diagonal dominance for low frequencies is to choose the matrix $\boldsymbol{K}_a$ as the inverse of the steady-state plant gain matrix. More sophisticated algorithms are based on optimization or pseudo-diagonalization methods. Those methods are often used as a starting point for future steps, which are made interactively with the Nyquist array graphs and Gershgorin bounds.

If diagonal dominance is achieved and the Gershgorin bounds are small, then in order to meet design requirements, SISO controllers may be designed separately for each control loop using well-established classical methods.

The direct Nyquist array method provides simultaneous insight into the structure, interaction and stability.

2.8 LINEAR/QUADRATIC CONTROL

2.8.1 Linear/quadratic control problem

Consider a linear, multivariable system represented by the state-space equations:

$$\begin{aligned} \dot{\boldsymbol{x}}(t) &= \boldsymbol{A}\boldsymbol{x}(t) + \boldsymbol{B}\boldsymbol{u}(t) & (2.120)\\ \boldsymbol{y}(t) &= \boldsymbol{C}\boldsymbol{x}(t). & (2.121) \end{aligned}$$

The linear/quadratic (LQ) problem aims to minimize the weighted sum of the energy of the state and control of the form of the cost function:

$$J = \frac{1}{2}\int_0^T \left(\boldsymbol{x}^T(t)\boldsymbol{Q}\boldsymbol{x}(t) + \boldsymbol{u}^T(t)\boldsymbol{R}\boldsymbol{u}(t)\right) dt \tag{2.122}$$

with respect to the input vector $\boldsymbol{u}(t)$. For the optimization, $\boldsymbol{Q}$ must be symmetric positive semidefinite ($\boldsymbol{Q}^T = \boldsymbol{Q} \geq \mathrm{o}$), and $\boldsymbol{R}$ must be symmetric positive definite ($\boldsymbol{R}^T = \boldsymbol{R} > \mathrm{o}$). Using, for instance, the Pontryagin minimum principle, the solution is of the form of a time-varying control law:

$$\boldsymbol{u}(t) = -\boldsymbol{K}(t)\boldsymbol{x}(t), \tag{2.123}$$

where:

$$\boldsymbol{K}(t) = \boldsymbol{R}^{-1}\boldsymbol{B}^T\boldsymbol{P}(t). \tag{2.124}$$

$\boldsymbol{P}(t)$ is a solution of the Riccati differential equation:

$$\boldsymbol{A}^T\boldsymbol{P}(t) + \boldsymbol{P}(t)\boldsymbol{A} + \boldsymbol{Q} - \boldsymbol{P}(t)\boldsymbol{B}\boldsymbol{R}^{-1}\boldsymbol{B}^T\boldsymbol{P}(t) = -\frac{d}{dt}\boldsymbol{P}(t). \tag{2.125}$$

To implement the LQ controller (2.123) all states $\boldsymbol{x}(t)$ are required to be measurable.

If the time horizon is infinite ($T \to \infty$) and an optimal solution exists, then $\boldsymbol{P}(t)$ tends to a constant matrix $\boldsymbol{P}$. The control law is:

$$\boldsymbol{u}(t) = -\boldsymbol{K}\boldsymbol{x}(t), \tag{2.126}$$

$$\boldsymbol{K} = \boldsymbol{R}^{-1}\boldsymbol{B}^T\boldsymbol{P}, \tag{2.127}$$

where $\boldsymbol{P}$ is a solution of the Riccati algebraic equation:

$$\boldsymbol{A}^T\boldsymbol{P} + \boldsymbol{P}\boldsymbol{A} + \boldsymbol{Q} - \boldsymbol{P}\boldsymbol{B}\boldsymbol{R}^{-1}\boldsymbol{B}^T\boldsymbol{P} = \mathrm{o}. \tag{2.128}$$

For the positive-definite matrix $\boldsymbol{P}$, the corresponding closed-loop system with the LQ controller (Fig. 2.16) is asymptotically stable.

LQ designed controllers ensure good stability margin and sensitivity properties. In the SISO case, the LQ control design technique enables gain margins in the range $[-6, \infty)$ dB to be obtained. A consequence of infinite gain margin is the minimum phase property of the open-loop transfer function. The corresponding phase margin is not less than $\frac{\pi}{3}$ [rad]. The sensitivity function is always less then 1. This implies good disturbance rejection and tracking properties.

2.8.2 Linear/quadratic Gaussian control

Suppose that a plant model is of the state-space form:

$$\dot{\boldsymbol{x}}(t) = \boldsymbol{A}\boldsymbol{x}(t) + \boldsymbol{B}\boldsymbol{u}(t) + \boldsymbol{\Gamma}\boldsymbol{w}(t), \tag{2.129}$$

$$\boldsymbol{y}(t) = \boldsymbol{C}\boldsymbol{x}(t) + \boldsymbol{v}(t), \tag{2.130}$$

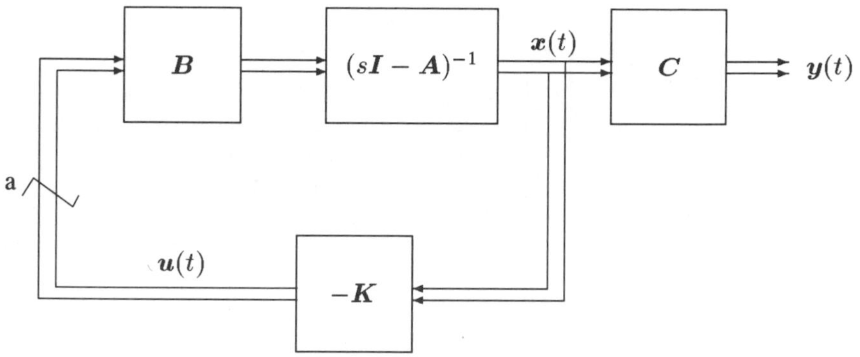

Figure 2.16. Linear/quadratic control problem

where $\boldsymbol{w}(t)$ and $\boldsymbol{v}(t)$ represent statistical knowledge about plant disturbances. They are zero mean Gaussian white noises with covariance matrices:

$$\mathrm{E}\{\boldsymbol{w}(t)\boldsymbol{w}^T(t)\} = \boldsymbol{W} \geq \mathbf{o}, \tag{2.131}$$

$$\mathrm{E}\{\boldsymbol{v}(t)\boldsymbol{v}^T(t)\} = \boldsymbol{V} > \mathbf{o}, \tag{2.132}$$

and uncorrelated:

$$\mathrm{E}\{\boldsymbol{v}(t)\boldsymbol{w}^T(t+\tau)\} = \mathbf{o}, \tag{2.133}$$

for all t and τ. $\mathrm{E}\{\cdot\}$ is the statistical expectation operator.

The linear/quadratic Gaussian (LQG) control problem is to devise a control law that minimizes the cost function:

$$J = \frac{1}{2} \lim_{T\to\infty} \mathrm{E}\left\{ \int_0^T \left(\boldsymbol{x}^T(t)\boldsymbol{Q}\boldsymbol{x}(t) + \boldsymbol{u}^T(t)\boldsymbol{R}\boldsymbol{u}(t)\right) dt \right\}. \tag{2.134}$$

Using the separation principle the LQG control problem can be decomposed into two subproblems:

- obtaining of an optimal estimate $\hat{\boldsymbol{x}}(t)$ of state $\boldsymbol{x}(t)$ minimizing

$$\mathrm{E}\left\{ \left(\boldsymbol{x}(t) - \hat{\boldsymbol{x}}(t)\right)^T \left(\boldsymbol{x}(t) - \hat{\boldsymbol{x}}(t)\right) \right\} \tag{2.135}$$

  using the Kalman filter theory. The optimal estimator is:

$$\dot{\hat{\boldsymbol{x}}}(t) = \boldsymbol{A}\hat{\boldsymbol{x}}(t) + \boldsymbol{B}\boldsymbol{u}(t) + \boldsymbol{L}(\boldsymbol{y}(t) - \boldsymbol{C}\hat{\boldsymbol{x}}(t)), \tag{2.136}$$

  where the filter gain $\boldsymbol{L}$ is

$$\boldsymbol{L} = \mathcal{P}\boldsymbol{C}^T\boldsymbol{V}^{-1}. \tag{2.137}$$

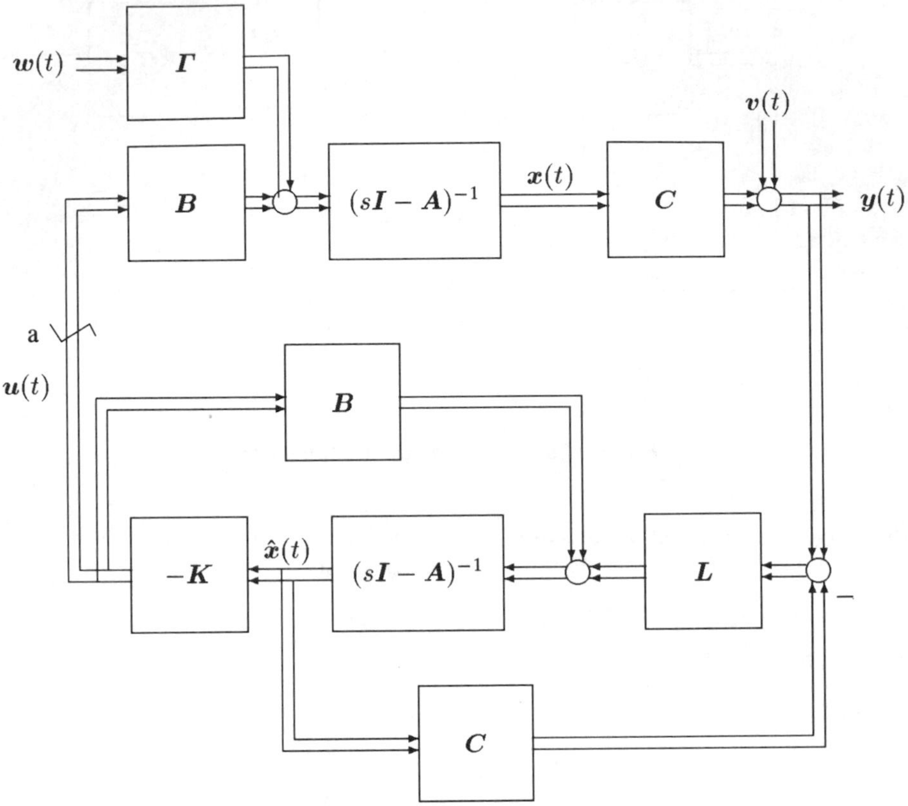

Figure 2.17. Linear/quadratic Gaussian control problem

The estimation error variance $\mathcal{P}$ is a solution of the filter Riccati algebraic equation:

$$A\mathcal{P} + \mathcal{P}A^T + \Gamma W \Gamma^T - \mathcal{P}C^T V^{-1} C \mathcal{P} = o; \tag{2.138}$$

- application of the standard deterministic LQ control law with the Kalman state estimates $\hat{x}(t)$ instead of state $x(t)$:

$$u(t) = -K\hat{x}(t)V^{-1}. \tag{2.139}$$

The block diagram of the LQG control problem is represented in Fig. 2.17.

The LQG cost function can be interpreted as a weighted H_2-norm of states and control signals. For instance in the SISO case, a controller $K_r(s)$ can be designed so that the error

$$||e(t)||_2 = \int_0^\infty e^2(t)dt \tag{2.140}$$

is minimized for a particular disturbance $d(t)$ at the output. The signal $e(t)$ is related to $d(t)$ by the transfer function, which is the sensitivity function $S(s)$. From the Parseval theorem it follows that:

$$\int_0^\infty e^2(t)dt = \frac{1}{2\pi}\int_{-\infty}^{\infty} |E(j\omega)|^2\, d\omega = \frac{1}{2\pi}\int_{-\infty}^{\infty} |S(j\omega)D(j\omega)|^2\, d\omega. \tag{2.141}$$

It follows that the LQG control problem can be interpreted as a minimization of the H_2-norm of the sensitivity function weighted by signal $D(j\omega)$.

2.8.3 Loop transfer recovery

Linear/quadratic controllers have good stability and sensitivity properties. There are difficulties with their real-life implementations because they require that all states be measurable.

When the closed loop on the block diagram (Fig. 2.16)) is broken at the point '**a**', the open-loop transfer function is

$$\boldsymbol{K}\,(s\boldsymbol{I}-\boldsymbol{A})^{-1}\,\boldsymbol{B}. \tag{2.142}$$

The corresponding open-loop transfer function for the LQG control problem (Fig. 2.17) is:

$$\boldsymbol{K}(s\boldsymbol{I}-\boldsymbol{A}+\boldsymbol{B}\boldsymbol{K}+\boldsymbol{L}\boldsymbol{C})^{-1}\boldsymbol{L}\boldsymbol{C}(s\boldsymbol{I}-\boldsymbol{A})^{-1}\boldsymbol{B}. \tag{2.143}$$

When the plant (2.129) is a minimum phase and the cost function J weights are chosen so that $\boldsymbol{R} = \mathrm{r}\boldsymbol{I}$ and $\boldsymbol{Q} = q^2\boldsymbol{B}\boldsymbol{B}^T$, we can show that for $q \rightarrow \infty$ the open-loop transfer function for the LQG control problem approaches that for the LQ control problem:

$$\lim_{q\rightarrow\infty}\left\{\boldsymbol{K}(s\boldsymbol{I}-\boldsymbol{A}|!+\boldsymbol{B}\boldsymbol{K}+\boldsymbol{L}\boldsymbol{C})^{-1}\boldsymbol{L}\boldsymbol{C}(s\boldsymbol{I}-\boldsymbol{A})^{-1}\boldsymbol{B}\right\} = \boldsymbol{K}\,(s\boldsymbol{I}-\boldsymbol{A})^{-1}\,\boldsymbol{B}. \tag{2.144}$$

This suggests a method of control system design which is known as loop transfer recovery. The control system requirements are expressed by a target feedback transfer function. These requirements can be met by the following two-step algorithm:

- Step 1 – loop shaping. For the matrix weight $\boldsymbol{Q} = \boldsymbol{C}^T\boldsymbol{C}$ the LQ controller is designed by varying the parameter r until the resulting open-loop transfer function is similar to the target feedback transfer function and the sensitivity and complementary sensitivity functions have desired shape;

- Step 2 – recovery. For increasing values of the parameter q the filter Riccati algebraic equation (2.138) is solved until the LQG control system approaches the LQ performance. The value of q should not be increased infinitely because it may lead to a physically difficult implementation of the filter gain $\boldsymbol{L}$.

Loop transfer recovery can be regarded as a frequency-domain method because we use state-space representation as a tool to meet the frequency-domain control system design requirements.

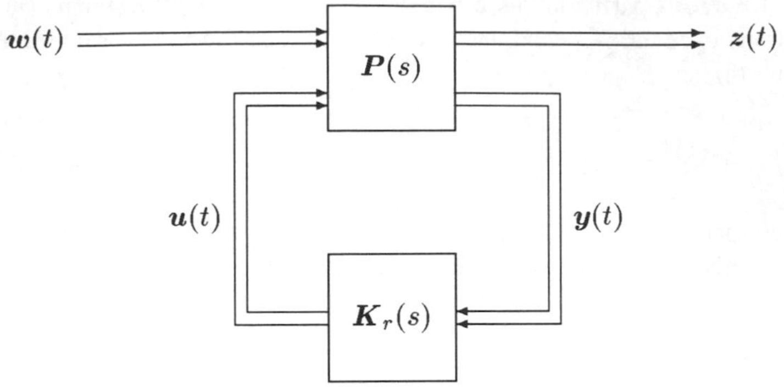

Figure 2.18. Two-port block diagram of the control problem

2.9 H_∞ CONTROL

2.9.1 Two-port control problem representation

A variety of control problems can be represented as a two-port diagram. It consists of a modified plant $\boldsymbol{P}(s)$ and controller $\boldsymbol{K}_r(s)$ blocks (Fig. 2.18). Plant inputs are grouped into:

- the vector of control input signals $\boldsymbol{u}(t)$;
- the vector of exogenous input signals $\boldsymbol{w}(t)$, i.e. the vector of signals which cannot be manipulated by the controller.

Plant outputs are also divided into two vectors:

- the vector $\boldsymbol{y}(t)$ which consists of signals that are measured and used as the input vector of the controller to produce the control signal $\boldsymbol{u}(t)$;
- the vector $\boldsymbol{z}(t)$ which is a set of signals used in measures of the closed-loop system performance.

Suppose that $\boldsymbol{P}(s)$ is partitioned as

$$\boldsymbol{P}(s) = \begin{bmatrix} \boldsymbol{P}_{11}(s) & \boldsymbol{P}_{12}(s) \\ \boldsymbol{P}_{21}(s) & \boldsymbol{P}_{22}(s) \end{bmatrix} \tag{2.145}$$

so that the two-port system equations are

$$\boldsymbol{Z}(s) = \boldsymbol{P}_{11}(s)\boldsymbol{W}(s) + \boldsymbol{P}_{12}(s)\boldsymbol{U}(s), \tag{2.146}$$

$$\boldsymbol{Y}(s) = \boldsymbol{P}_{21}(s)\boldsymbol{W}(s) + \boldsymbol{P}_{22}(s)\boldsymbol{U}(s), \tag{2.147}$$

$$\boldsymbol{U}(s) = \boldsymbol{K}_r(s)\boldsymbol{Y}(s). \tag{2.148}$$

From these equations follows that the vector $\boldsymbol{z}(t)$ and vector $\boldsymbol{w}(t)$ are related by:

$$\boldsymbol{Z}(s) = \boldsymbol{F}(\boldsymbol{P}(s), \boldsymbol{K}_r(s))\boldsymbol{W}(s) \tag{2.149}$$

where

$$\boldsymbol{F}(\boldsymbol{P}(s), \boldsymbol{K}_r(s)) = \boldsymbol{P}_{11}(s) + \boldsymbol{P}_{12}(s)\boldsymbol{K}_r(s)\left(\boldsymbol{I}(s) - \boldsymbol{P}_{22}(s)\right)^{-1}\boldsymbol{P}_{21}(s) \tag{2.150}$$

is called the linear fractional transformation.

A suitable definition of signals $\boldsymbol{w}(t)$ and $\boldsymbol{z}(t)$ or equivalently the corresponding transfer-function matrix $\boldsymbol{P}(s)$ allows us to put a number of control system design problems into two-port representations. Different measures of $\boldsymbol{F}(\boldsymbol{P}(s), \boldsymbol{K}_r(s))$ can be used to describe the desired control system performance.

The plant can also be represented in the state-space form:

$$\dot{\boldsymbol{x}}(t) = \boldsymbol{A}\boldsymbol{x}(t) + \boldsymbol{B}_1\boldsymbol{w}(t) + \boldsymbol{B}_2\boldsymbol{u}(t), \tag{2.151}$$

$$\boldsymbol{z}(t) = \boldsymbol{C}_1\boldsymbol{x}(t) + \boldsymbol{D}_{11}\boldsymbol{w}(t) + \boldsymbol{D}_{12}\boldsymbol{u}(t), \tag{2.152}$$

$$\boldsymbol{y}(t) = \boldsymbol{C}_2\boldsymbol{x}(t) + \boldsymbol{D}_{21}\boldsymbol{w}(t) + \boldsymbol{D}_{22}\boldsymbol{u}(t). \tag{2.153}$$

This implies the following packed-form matrix notation:

$$\boldsymbol{P}(s) = \begin{bmatrix} \boldsymbol{A} & \boldsymbol{B}_1 & \boldsymbol{B}_2 \\ \boldsymbol{C}_1 & \boldsymbol{D}_{11} & \boldsymbol{D}_{12} \\ \boldsymbol{C}_2 & \boldsymbol{D}_{21} & \boldsymbol{D}_{22} \end{bmatrix}. \tag{2.154}$$

2.9.2 H_∞ control problem

Consider a two-port control problem representation (Fig. 2.18). The objective of H_∞ control is to find an internally stabilizing and realizable controller $\boldsymbol{K}_r(s)$ for a given plant $\boldsymbol{P}(s)$ such that the H_∞-norm of the linear fractional transformation matrix $\boldsymbol{F}(\boldsymbol{P}(s), \boldsymbol{K}_r(s))$ is below a given level γ ($\gamma \in \mathcal{R}$ and $\gamma > 0$):

$$||\boldsymbol{F}(\boldsymbol{P}(s), \boldsymbol{K}_r(s))||_\infty < \gamma, \tag{2.155}$$

which is the standard H_∞ control problem. The aim of an optimal problem is to find a controller which minimizes:

$$||\boldsymbol{F}(\boldsymbol{P}(s), \boldsymbol{K}_r(s))||_\infty. \tag{2.156}$$

The solution of the optimal H_∞ control problem can be obtained iteratively by solving the standard problem for growing smaller values of γ until the solution does not exist. If the value of γ is increased to a very large value, then the algorithm converges to the LQG controller which minimizes:

$$||\boldsymbol{F}(\boldsymbol{P}(s), \boldsymbol{K}_r(s))||_2. \tag{2.157}$$

An initial value of γ can be chosen as the peak of the closed-loop transfer function gain obtained from a solution of the LQG control problem. This algorithm is called the γ iteration.

The most often used optimal H_∞ control problems are the following:

- the sensitivity minimization problem with the cost function:

$$\boldsymbol{F}(\boldsymbol{P}(s), \boldsymbol{K}_r(s)) = \boldsymbol{W}_S(s)\boldsymbol{S}(s) = \boldsymbol{W}_S(s)\left(\boldsymbol{I} - \boldsymbol{K}_o(s)\boldsymbol{K}_r(s)\right)^{-1}. \quad (2.158)$$

Its two-port representation is:

$$\boldsymbol{P}_{11}(s) = \boldsymbol{W}_S(s), \quad (2.159)$$
$$\boldsymbol{P}_{12}(s) = \boldsymbol{W}_S(s)\boldsymbol{K}_r(s), \quad (2.160)$$
$$\boldsymbol{P}_{21}(s) = \boldsymbol{I}, \quad (2.161)$$
$$\boldsymbol{P}_{22}(s) = -\boldsymbol{K}_o(s). \quad (2.162)$$

- the mixed robust stability and robust performance problem with the cost function:

$$\boldsymbol{F}(\boldsymbol{P}(s), \boldsymbol{K}_r(s)) = \begin{bmatrix} \boldsymbol{W}_S(s)\boldsymbol{S}(s) \\ \boldsymbol{M}(s)\boldsymbol{\Delta}_{mo}(s) \end{bmatrix}. \quad (2.163)$$

Its two-port representation is:

$$\boldsymbol{P}_{11}(s) = \begin{bmatrix} \boldsymbol{W}_S(s) \\ 0 \end{bmatrix}, \quad (2.164)$$
$$\boldsymbol{P}_{12}(s) = \begin{bmatrix} -\boldsymbol{W}_S(s)\boldsymbol{K}_o(s) \\ \boldsymbol{\Delta}_{mo}(s)\boldsymbol{K}_o(s) \end{bmatrix}, \quad (2.165)$$
$$\boldsymbol{P}_{21}(s) = \boldsymbol{I}, \quad (2.166)$$
$$\boldsymbol{P}_{22}(s) = -\boldsymbol{K}_o(s). \quad (2.167)$$

- the mixed sensitivity problem with the cost function:

$$\boldsymbol{F}(\boldsymbol{P}(s), \boldsymbol{K}_r(s)) = \begin{bmatrix} \boldsymbol{W}_S(s)\boldsymbol{S}(s) \\ \boldsymbol{W}_M(s)\boldsymbol{M}(s) \end{bmatrix}. \quad (2.168)$$

The two-port representation of this problem is easy to obtain using the previous representations.

2.10 LABORATORY EXERCISES

2.10.1 MATLAB software tools applied in laboratory course 2

In the last decade, classical and modern control system design methods involving advanced mathematical techniques and time-consuming calculations have been aided by software packages like MATLAB and SIMULINK with toolboxes. These packages automate the calculations enabling the designer to concentrate on art of control system design.

The MATLAB Control System Toolbox provides a collection of the basic and most popular tools for control system design. The capabilities of the Control System Toolbox are extended by the Multivariable Frequency Domain Toolbox. Robust feedback control analysis and synthesis methods are given in the Robust Control Toolbox. Those toolboxes provide a wide range of model representations, methods for system validation in the time and frequency domains and control algorithms. Those tools can be used for creating customized applications by writing new m-files.

Some of functions of the Control System Toolbox, Multivariable Frequency Domain Toolbox and Robust Control Toolbox are used in this laboratory course. Exercises are illustrated with simulations performed with SIMULINK.

A selection of basic commands is included in each example. The detailed description of the toolboxes can be found in the user guides.

2.10.2 SISO plant representation

Example 1 – frequency response

Consider the following transfer function of a plant:

(a) $$K_o(s) = \frac{1}{sT+1} \tag{2.169}$$

where $T > 0$;

(b) $$K_o(s) = \frac{\omega_n^2}{s^2 + 2\xi\omega_n s + 1} \tag{2.170}$$

where $\omega_n > 0$ and $0 \leq \xi < 1$.

Plot the Nyquist and Bode graphs for different values of plant parameters T, ω_n and ξ.

Quick help

1. The following commands declare vectors of the first-order inertia parameters ($T = 3$):

```
>> numo = 1;
>> deno = [3,1];
```

2. The values of a frequency scale vector `w` are generated by the `linspace` or `logspace` commands. The syntax is as follows:

```
w = linspace(x,y,z);
w = logspace(x,y,z);
```

Those commands produce linearly (from `x` to `y`) or logarithmically (from 10^x to 10^y) spaced values of the frequency scale. The number `z` of points is optional.

3. The frequency response of a system can be obtained using the `bode` or `nyquist` commands:

`>> bode(num,den,w)` produces, on a split screen, the Bode plots;
`>> nyquist(num,den,w)` computes and plots the Nyquist plot of the frequency response.

The frequency scale is `w`. Try also:

```
>> bode(num,den);
>> [magnitude,phase, w] = bode(num,den);
>> [magnitude,phase] = bode(num,den,w);
```

Additional useful commands are `semilogx` and `semilogy`. Check their significance using the MATLAB help.

Questions

1. Find expressions for the magnitude and phase of the frequency response for given plants.
2. How does the maximum value of the frequency response magnitude vary with the changes of ω_n and ξ?

Example 2 – uncertainty models and system measures

The model of the plant:

(a)
$$K_o(s) = \frac{1}{(sT_1+1)(sT_2+1)} \tag{2.171}$$
where $T_1, T_2 > 0$

(b)
$$K_o(s) = \frac{\omega_n^2}{s^2+2\xi\omega_n s+1} \tag{2.172}$$
where $\omega_n > 0$ and $0 \le \xi < 1$

for the control system design purpose is chosen as:

$$\tilde{K}_o(s) = \frac{1}{sT+1}. \tag{2.173}$$

Find the corresponding additive and multiplicative uncertainty models. Plot their Nyquist and Bode magnitude graphs. Calculate the H_2- and H_∞-norms of the plant and uncertainty transfer functions for different values of parameters T_1, T_2, ω_n, ξ and T.

Quick help

1. Using the SIMULINK platform, simulate the plant excited by white noise with variance equal to 1 (the `csd_1` program). The H_2-norm of the plant transfer function can be obtained as the mean square value of the plant output signal.
2. The H_∞-norm of a plant transfer function can calculated as the maximum value of the transfer-function magnitude:

   ```
   hinfty = max(magnitude);
   ```

   where the vector `magnitude` is produced by the `bode` command. The frequency scale vector `w` (the input parameter of the `bode` command) should be chosen to cover all dynamical modes. Values of the H_2- and H_∞-norm can be calculated by using `normh2` and `normhinf` commands from the Robust Control Toolbox. Check their synopsis using help.

Questions

1. Calculate the H_2- and H_∞-norms of the plant and its uncertainty models without using MATLAB. Compare the results with these obtained using SIMULINK and MATLAB. Explain differences.
2. How does the choice of the frequency scale vector w influence the results of the H_∞-norm calculation for the different values of plant transfer-function parameters ω_n and ξ?

2.10.3 Feedback systems

Example 1 – stability

Consider the following transfer function of a plant:

$$K_o(s) = \frac{1}{s^2 + 4s + 3} \tag{2.174}$$

and a rational feedback of the form:

$$K_r(s) = k\frac{s+p}{s}. \tag{2.175}$$

Plot the Nyquist and Bode graphs of the open-loop frequency response for different values of feedback parameters k and p.

Quick help

1. The following commands declare vectors of the plant and feedback transfer-functions parameters:

```
>> numo = 1;
>> deno = [1,4,3];
>> numr = [k,kp];
>> denr = [1,0];
```

2. The open-loop transfer function $K_o(s)K_r(s)$ can be obtained using the `series` command:

```
>> [num,den] = series(numo,deno,numr,denr);
```

See also `parallel`, `feedback` and `cloop`.

3. The gain and phase margins may be calculated using the `margin` command:

```
>> [gm,pm,wp,wg] = margin(mag,phase,w);
```

where the transfer-function magnitude (`mag`), phase (`phase`) and frequency range(w) are vectors produced by the `bode` command:

```
>> [mag,phase,w] = bode(num,den);
```

Try also:

```
>> margin(mag,phase,w);
```

Questions

1. Write the characteristic equation of the closed-loop system.
2. Use the Hurwitz or Routh test to determine for what values of k and p the system will be stable. Check your answer with Nyquist or Bode plots.
3. Obtain the gain and phase margins and the corresponding gain crossover frequency for different values of stable feedback parameters k and b. How do values of k and p affect the stability margin?
4. For a given value of p find the value of k for a critically damped response.
5. Check the internal stability of the closed-loop system.
6. Find frequency-dependent upper bounds on the magnitude of the additive and multiplicative uncertainty that ensure robust stability of the closed-loop system.
7. Calculate the additive and multiplicative robust stability margins for different values of stable feedback parameters k and b.

Example 2 – sensitivity and complementary sensitivity

For the control system from example 1, plot graphs of the sensitivity and complementary sensitivity functions for different values of feedback parameters k and p.

Quick help

1. The closed-loop transfer function can be obtained using the `cloop` command:

   ```
   >> [clnum,clden] = cloop(num,den,-1);
   ```

   where `-1` denotes negative-feedback.
2. The complementary sensitivity function is the closed-loop transfer gain function calculated above.
3. The sensitivity function can be obtained as the inverse of the transfer-function magnitude of $\frac{nums(s)}{dens(s)}$:

   ```
   >> [nums,dens] = parallel(1,1,num,den);
   ```

Questions

1. Discuss the influence of the feedback transfer-function parameterization ($K_r(s)$) on the sensitivity and complementary sensitivity functions.
2. How do the resonance frequency and corresponding maximum value vary with the changes of k or p?
3. How does k or p affect the disturbance stop band (or the pass band)?
4. Which values of k and p give zero steady-state error? Why?

Example 3 – time responses

Using SIMULINK, simulate the plant and closed-loop system from example 1. Use as input signal a sine wave or unit step. Repeat the simulations for different values of feedback parameters k and p .

Quick help

1. In the MATLAB and SIMULINK environment enter the program `csd_2`. This program enables the complementary sensitivity function measurements. The frequency of the input sine wave should be varied to obtain values of the complementary sensitivity function.
2. To perform step response experiments invoke the `csd_3` program.

Questions

1. How do the sensitivity function values measured with sine input under steady-state conditions correspond with the results obtained in example 1?
2. Design any experiment which allows you to measure the sensitivity function in SIMULINK. Perform some experiments and compare the results with example 1.
3. How do the values of k and p affect the steady-state error, overshoot, settling and rise time?

Example 4 – root locus

Obtain a root locus plot for the system from example 1 for a chosen set of values p.

Quick help

1. The root locus plot can be generated using the `rlocus` command:

   ```
   >> rlocus(num,den);
   ```

   The feedback gain is adjusted automatically.
2. The gain (`k`) at a given point and the corresponding closed-loop poles (`poles`) on the root locus plot can be obtained interactively using the mouse and the command:

   ```
   >> [k,poles] = rlocfind(num,den);
   ```

   It is important to plot the root locus plot first.

Questions

1. How does the controller pole described by parameter p affect the root locus and stability?
2. Find the value of k for a given p for a critically damped response. Compare the results with previous examples.

Example 5 – robust stability and performance

For plants from example 1 (subsection **SISO plant representation**) check the robust stability using the additive and multiplicative uncertainty models. Plot graphs of the sensitivity and complementary sensitivity for perturbed plants.

A designed control system should reduce the sensitivity at least as $50 : 1$ up to the frequency 2 (4 or 5) $\left[\frac{\text{rad}}{\text{sec}}\right]$. The corresponding closed-loop pass band should be up to 30 (40 or 60) $\left[\frac{\text{rad}}{\text{sec}}\right]$, respectively.

Quick help

1. Try to declare frequency-dependent bounds on the sensitivity ($W_S(j\omega)$) and complementary sensitivity ($W_M(j\omega)$) functions. Approximate these bounds by using the transfer function:

$$W(s) = k_w \frac{sT_1 + 1}{sT_2 + 1}; \tag{2.176}$$

where:

- for the sensitivity function use $T_1 \ll T_2$,
- for the complementary sensitivity function use $T_1 \gg T_2$.

Choose proper values of parameters k_w, T_1 and T_2 to approximate these bounds.

Questions

1. Are the robust stability and robust performance achieved simultaneously? Why?
2. Calculate bounds on the open-loop transfer-function magnitude for chosen transfer-function bounds $W_S(s)$ nd $W_M(s)$.

2.10.4 Classical system design

Example 1 – PID control

For the following plants:

(a) $$K_o(s) = \frac{1}{s(s+4)(s+8)}; \tag{2.177}$$

(b) $$K_o(s) = \frac{1}{s(s+2)} \tag{2.178}$$

design a P, PI or PID controller using:

1. the stability margin specifications;
2. the M_{max} criterion;
3. the Ziegler–Nichols controller settings.

Quick help

1. A step response of a system described by its transfer function is generated by using the `step` command. Check its syntax using the MATLAB help.

2. The poles and zeros of a transfer function $\frac{num(s)}{den(s)}$ can be obtained with the `tf2zp` command:

```
>> [zeros,poles,gain] = tf2zp(num,den);
```

Questions

1. Compare the closed-loop poles and zeros, the phase and gain margins, and step responses of the closed-loop system for feedbacks obtained using different design principles.

2. Modify the controller settings to meet additional design specifications:

   - overshoot $< 15\%$ (or 35%) or
   - settling time $< 1s$ (or $15s$).

   Compare the gain and phase margins, and step responses of the closed-loop system for both cases.

Example 2 – pole placement

A plant is described by the state-space equations:

$$\dot{\boldsymbol{x}}(t) = \begin{bmatrix} -0.30 & 0.10 & -0.05 \\ 1.00 & 0.10 & 0.00 \\ -1.50 & -8.90 & -0.05 \end{bmatrix} \boldsymbol{x}(t) + \begin{bmatrix} 2.00 \\ 0.00 \\ 4.00 \end{bmatrix} u(t) \tag{2.179}$$

$$y(t) = \begin{bmatrix} 0.00 & 0.00 & 1.00 \end{bmatrix} \boldsymbol{x}(t). \tag{2.180}$$

Design state feedback to place the system poles at -5, $-2+j$ and $-2-j$.

Quick help

1. State-space matrices $\boldsymbol{A}, \boldsymbol{B}, \boldsymbol{C}$ and $\boldsymbol{D}$ are entered in the MATLAB environment using the commands:

```
   >> a = [-0.30,0.10,-0.5; 1.00,0.10,0.00;
-1.50,-8.90,-0.05];
   >> b = [2.00; 0.00; 4.00];
   >> c = [0.00,0.00,1.00];
   >> d = 0;
```

2. The state-space model can be converted into poles and zeros by using the `ss2zp` command:

```
[zeros,poles,gain] = ss2zp(a,b,c,d);
```

3. A transfer function $\frac{num(s)}{den(s)}$ for a plant described by state-space equations can be calculated by using the `ss2tf` command:

```
>> [num,den] = ss2tf(a,b,c,d);
```

4. The state-space feedback gain vector `k` can be obtained using the `place` command:

```
k = place(a,b,[-5,-2+j],[-2-j]);
```

5. The Nyquist and Bode plots for a state-space model can be generated using for instance commands:

```
>> bode(a,b,c,d);
>> nyquist(a,b,c,d);
```

Questions

1. Check the locations of plant poles and zeros.
2. Try to calculate the gain vector `k` without using MATLAB.
3. Obtain the transfer function of the compensated plant.
4. Compare the Nyquist and Bode plots, gain and phase margins and step responses of the original and compensated plant.

2.10.5 Multivariable control systems

Example 1 – control structure design

Suppose a three-input, three-output system is described by its steady-state gain matrix:

$$\begin{bmatrix} y_1 \\ y_2 \\ y_3 \end{bmatrix} = \begin{bmatrix} 3.90 & 0.17 & 4.00 \\ -1.30 & -0.20 & 0.00 \\ -3.10 & 1.00 & -2.40 \end{bmatrix} \begin{bmatrix} u_1 \\ u_2 \\ u_3 \end{bmatrix} \tag{2.181}$$

Calculate the relative gain array, Niederliński indexes and condition numbers for all possible input–output pairings.

Quick help

1. The steady-state gain matrix is entered in MATLAB in the following way:

```
>> k0 = [3.90 0.17 4.00; -1.30 -0.20 0.00; -3.10 1.00
-2.40];
```

2. The relative gain array ($\boldsymbol{rga}$) can be calculated using the `inv` command:

```
>> rga = k0.*inv(k0)';
```

3. The Niederliński index for a chosen pairing can be obtained using the syntax:

```
ni = det(k0)/(k0(1,1)*k0(2,2)*k0(3,3));
```

Questions

1. How can inputs and outputs be paired?
2. Which input–output pairings result in an unstable closed-loop system?
3. For a chosen set of possible input–output pairings compare condition numbers. Is the plant easy to control?

Example 2 – multivariable system stability

Consider a multivariable plant described by the transfer function matrix:

$$\boldsymbol{K}_o(s) = \begin{bmatrix} \frac{e^{-2s}}{10s+1} & \frac{1}{10s+1} \\ \frac{0.1}{5s+1} & \frac{e^{-10s}}{60s+1} \end{bmatrix} \tag{2.182}$$

Check the closed-loop system stability using the characteristic loci and Nyquist array with the Gershgorin circles assuming a decentralized proportional controller:

$$\boldsymbol{K}_r(s) = \begin{bmatrix} k_1 & 0 \\ 0 & k_2 \end{bmatrix} \tag{2.183}$$

for different values of k_1 and k_2.

Quick help

1. The plant and controller transfer-function matrix parameters are declared using a common denominator polynomial form of $\boldsymbol{K}_o(s)$ and $\boldsymbol{K}_r(s)$:

```
>> kr = [k1 0; 0 k2];
>> denp = [3000 900 65 1];
>> nump = [0 300 65 1 0 300 65 1; 0 600 70 1 0 50 15
1];
>> dl = [2 0; 0 10];
```

2. The multivariable frequency response matrix of the plant can be calculated by adding delays (the `fdly` command) to the multivariable frequency response obtained by using the `mv2fr` command:

```
>> f1 = mv2fr(nump,denp,w);
>> fp = fdly(w,f1,dl);
```

The frequency vector `w` is not generated automatically and should be specified before involving the above commands.

3. The multivariable frequency response of a cascaded proportional controller $\boldsymbol{K}_r(s)$ and plant can be calculated using the `fmul` command:

```
>> f=fmul(w,fp,kr);
```

4. The characteristic loci of the multivariable frequency response `f` is generated by use of the `feig` command:

```
>> cl = feig(w,f);
```

The characteristic loci can be plotted by using the `plotnyq` command:

```
>> plotnyq(csort(cl));
```

The `csort` command sorts columns of eigenvectors into continuous vectors.

5. The graph of the element (i, j) of the Nyquist array can be generated using the `plotnyq` command:

```
>> plotnyq(fget(w,fp,[i,j]));
```

where the command `fget` gets elements from the multivariable frequency response matrix.

6. The following set of commands puts the Gershgorin circles on the Nyquist plot of the element (i, j) of the multivariable frequency response matrix:

```
>> plotnyq(fget(w,fp,[i,j]));
>> hold on
>> plotnyq(fcgersh(w,f,i));
```

where the last command generates the Gershgorin circles.

Questions

1. Write the characteristic equation for the closed-loop system.
2. Find values of k_1 and k_2 that ensure critically damped closed-loop responses.
3. Is the plant transfer-function matrix diagonally dominant? Why?

Example 3 – multivariable control system design

For the plant from the previous example, design a multivariable controller using:

1. the sequential loop closing method;
2. the Nyquist array method;
3. the characteristic locus method.

Quick help

1. A permutation matrix K_a that assures diagonal dominance for a given frequency `w(i)` can be obtained using the `align` command:

   ```
   >> ka = align(inv(fgetf(w,fd,i)));
   ```

   The command `fgetf` returns components of the multivariable frequency response matrix for the frequency `w(i)`. See also more advanced methods for obtaining diagonal dominance: `fcnmat`, `frnmat`, `fperron`, `fpseudo` and `fadj`.

2. The multivariable frequency response matrix of a series connection of the plant ($nump$, $denp$) and general controller ($numc$, $denc$) can be generated by the following set of commands:

   ```
   >> [ap,bp,cp,dp] = mvtf2ss(nump,denp);
   >> [ac,bc,cc,dc] = mvtf2ss(numc,denc);
   >> [a,b,c,d] = mser(ap,bp,cp,dp,ac,bc,cc,dc);
   >> f = mv2fr(a,b,c,d,w);
   ```

   where:

   - `mvtf2ss` – converts the multivariable frequency transfer-function matrix to a state-space model;
   - `mvser` – generates the state-space model for the series connection of two dynamical systems.

3. The closed-loop state-space model can be calculated using the following commands:

   ```
   >> [as,bs,cs,ds] = mvtf2ss(eye(2),1);
   >> [al,bl,cl,dl] = mvfb(a,b,c,d,as,bs,cs,ds);
   ```

   where the `mvtf` command generates a state-space description of the multivariable feedback system.

4. The characteristic loci of the frequency response matrix can be manipulated by a compensator designed with the `facc` command. Details and examples are presented in the Multivariable Frequency Domain Toolbox manual.

5. Singular values of the sensitivity and complementary sensitivity functions can be obtained by using the `fsvd` command for multivariable frequency response matrices calculated with the `frd` (sensitivity function) and `ffb` (complementary sensitivity function) commands, respectively. Check their syntax using help.

Questions

1. How does the permutation matrix K_a obtained with the `align` command affect the diagonal dominance for different frequencies `w(i)`?

2. Compare results of all designs taking into account singular values of the sensitivity and complementary sensitivity functions, and step responses.

3. What are steady-state errors for all control loops?
4. Compare the additive and multiplicative robust stability margins.

2.10.6 Linear/quadratic control

Example 1 – linear/quadratic controller

For a plant described by:

$$\dot{\boldsymbol{x}}(t) = \begin{bmatrix} -0.30 & 0.10 & -0.05 \\ 1.00 & 0.10 & 0.00 \\ -1.50 & -8.90 & -0.05 \end{bmatrix} \boldsymbol{x}(t) + \begin{bmatrix} 2.00 \\ 0.00 \\ 4.00 \end{bmatrix} u(t) \tag{2.184}$$

$$y(t) = \begin{bmatrix} 0.00 & 0.00 & 1.00 \end{bmatrix} \boldsymbol{x}(t). \tag{2.185}$$

design a linear/quadratic controller that minimizes the cost function:

$$J = \frac{1}{2}\int_0^\infty \left(q\boldsymbol{x}^T(t)\boldsymbol{x}(t) + ru^2(t)\right) dt \tag{2.186}$$

Perform the design for different values of q and r.

Quick help

1. The command `lqr` solves the linear/quadratic control problem:

```
>> [k,p,e] = lqr(a,b,g*eye(3),r);
```

where `k` is the optimal feedback gain matrix, `p` is the unique positive-definite solution of the associated Riccati matrix equation and `e` is the vector of the closed-loop eigenvalues. Matrices `a` and `b` are declared as:

```
>> a = [-0.30,0.10,-0.5; 1.00,0.10,0.00;
-1.50,-8.90,-0.05];
>> b = [2.00; 0.00; 4.00];
```

See also the `are` command.

Questions

1. Check the stability of the closed-loop system with a linear/quadratic controller for all design cases.
2. Obtain the open-loop transfer function to calculate the gain and phase margins. Compare the results for different values of q and r.
3. Compare the sensitivity of pole and zero locations and step responses of the closed-loop system, for changing values of q and r.
4. Compare the additive and multiplicative robust stability margins for different values of q and r. Comment on the robustness of LQ designed controllers.

Example 2 – linear/quadratic Gaussian control

For the following plant:

$$\dot{\boldsymbol{x}}(t) = \begin{bmatrix} -0.30 & 0.10 & -0.05 \\ 1.00 & 0.10 & 0.00 \\ -1.50 & -8.90 & -0.05 \end{bmatrix} \boldsymbol{x}(t) + \begin{bmatrix} 2.00 \\ 0.00 \\ 4.00 \end{bmatrix} u(t) + \begin{bmatrix} 0.00 \\ 0.00 \\ 1.00 \end{bmatrix} w(t) \quad (2.187)$$

$$y(t) = \begin{bmatrix} 0.00 & 0.00 & 1.00 \end{bmatrix} \boldsymbol{x}(t) + v(t). \quad (2.188)$$

where:

$$\boldsymbol{W} = p\boldsymbol{I}, \quad (2.189)$$

$$\boldsymbol{V} = v, \quad (2.190)$$

design a LQG controller that minimizes:

$$J = \frac{1}{2}\int_0^\infty \left(q\boldsymbol{x}^T(t)\boldsymbol{x}(t) + ru^2(t)\right) dt. \quad (2.191)$$

Perform the design for different values of p, v, g and r.

Quick help

1. The gain matrix `l` of the Kalman filter, the corresponding error variance matrix `p` and the closed-loop eigenvalues `e` can be calculated as:

```
>> [l,p,e] = lqe(a,gamma,c,p*eye(3),v);
```

where:

```
>> gamma = [0;0;1];
>> a = [-0.30,0.10,-0.5; 1.00,0.10,0.00;
-1.50,-8.90,-0.05];
>> c = [0.00,0.00,1.00];
```

2. A solution of the loop transfer recovery control design problem can be found by using the `ltru` and `ltry` commands.

Questions

1. Compare the solutions of the Riccati filter equation and the optimal filter gain `l` for different values of parameters p and v.

2. Obtain the open-loop transfer function for the LQG optimal controller and different values of p and v. Calculate stability margins.

3. Compare properties of LQ controllers from the previous example with the LQG design.

4. Compare the sensitivity of pole and zero locations and step responses of the LQG closed-loop system, for changing values of q and r.

5. Compare the additive and multiplicative robust stability margins for different values of q and v. Comment on the robustness of LQG designed controllers.

6. Perform the loop transfer recovery design to obtain a controller that guarantees a chosen stability margin, e.g. gain margin equal to 6 dB and phase margin equal to $\frac{\pi}{4}$

2.10.7 H_∞ control

Example 1 – H_∞ control

Consider the following plants:

(a) $$K_o(s) = \frac{1}{(s+4)(s+8)}; \tag{2.192}$$

(b) $$K_o(s) = \frac{1}{(s+2)}; \tag{2.193}$$

(c) $$K_o(s) = \frac{1}{s(s+4)(s+8)}; \tag{2.194}$$

(d) $$K_o(s) = \frac{1}{s(s+2)}; \tag{2.195}$$

(e) $$\boldsymbol{K}_o(s) = \begin{bmatrix} \dfrac{e^{-2s}}{10s+1} & \dfrac{1}{10s+1} \\ \\ \dfrac{0.1}{5s+1} & \dfrac{e^{-10s}}{60s+1} \end{bmatrix}. \tag{2.196}$$

Design controllers which minimize:

1. sensitivity;
2. mixed robust stability and robust performance;
3. mixed sensitivity.

Quick help

1. The plant model for the H_∞-norm control problem design under the Robust Control Toolbox is declared by the `mksys` command. The plant (2.192) is inputted as:

```
>> [a,b,c,d] = tf2ss(1, [1 12 32]);
>> ssg = mksys(a,b,c,d);
```

For more details use help.

2. An augmented transfer-function model of the plant with weighting functions $\boldsymbol{W}_S(s)$, $\boldsymbol{W}_2(s)$ and $\boldsymbol{W}_M(s)$ for the general mixed sensitivity problem:

$$\left\| \begin{matrix} \boldsymbol{W}_S(s)\boldsymbol{S}(s) \\ \boldsymbol{W}_2(s)\boldsymbol{K}_r(s)\boldsymbol{S}(s) \\ \boldsymbol{W}_M(s)\boldsymbol{M}(s) \end{matrix} \right\|_\infty \tag{2.197}$$

can be computed by using the `augtf` command:

```
>> tss = augtf(ssg,ws,w2,wm);
```

where `ws`, `w2` and `wm` are matrices with declared transfer functions parameters that represent bounds on the sensitivity and complementary sensitivity functions. For instance, the transfer function:

$$W_S(s) = \frac{sT_1 + 1}{sT_2 + 1} \tag{2.198}$$

is declared as:

```
>> ws = [T1,1;T2,1];
```

See also `augss` and `augd`.

3. The H_∞-optimal controller synthesis by using the γ iteration algorithm is invoked by the `hinfopt` command:

```
>> [gammaopt,ssf,sscl] = hinfopt(tss);
```

`gammaopt` is the optimal value of γ. The variables `ssf` and `sscl` are packed representations of the controller and closed-loop transfer functions, respectively. They can be extracted by the `branch` command. See also the `hinf` and `linf` commands.

4. For plants with poles on the $j\omega$ axis use the `bilin` command to shift these poles into the left-half s-plane. Design a standard H_∞ controller for shifted problem and shift back the controller transfer function. Details can be found in the Robust Control Toolbox manual.

5. The H_2 control problem can be solved by the `h2lqg` command.

6. Time delay can be approximated by a rational transfer function generated by the command from the Control System Toolbox.

Questions

1. Compare closed-loop poles and zeros, the phase and gain margins, the robust additive and multiplicative stability margins, the sensitivity and complementary sensitivity functions, and step responses of the closed-loop system for feedbacks obtained using different H_∞ controller design objectives.

2. Discuss steady-state errors for all control design problems.

3. Compare the results obtained with those obtained in previous examples.

4. Repeat previous activities for the H_2-norm minimization control design problem.

2.11 REFERENCES

[1] B. D. O. Anderson and J. B. Moore. *Optimal Filtering.* Prentice Hall, Englewood Cliffs, NJ, 1979.

[2] B. D. O. Anderson and J. B. Moore. *Optimal Control.* Prentice Hall, Englewood Cliffs, NJ, 1990.

[3] K. J. Åström and B. Wittenmark. *Computer Controlled Systems: Theory and Design.* Prentice Hall, Englewood Cliffs, NJ, 1990.

[4] R. Y. Chiang and M. G. Safonov. *Robust Control Toolbox for Use with MATLAB. User's Guide.* MathWorks, 1992.

[5] C. A. Desoer and M. Vidyasagar. *Feedback Systems: Input-Output Properties.* Academic Press, New York, 1975.

[6] P. B. Desphande. *Multivariable Process Control.* Instrument Society of America, 1989.

[7] R. C. Dorf. *Modern Control Systems.* Addison-Wesley, Reading, MA, 1989.

[8] J. C. Doyle, B. A. Francis and A. R. Tannenbaum. *Feedback Control Theory.* Macmillan, New York, 1992.

[9] M. P. Ford, J. M. Maciejowski and J. M. Boyle. *Multivariable Frequency Domain Toolbox for Use with MATLAB. User's Guide.* MathWorks, 1992.

[10] G. F. Franklin, J. D. Powell and A. Emmani-Naeimi. *Feedback Control of Dynamic Systems.* Addison Wesley, Reading, MA, 1991.

[11] B. A. Francis. *A Course in H^∞ Control Theory.* Springer-Verlag, New York, 1987.

[12] A. Grace, A. J. Laub, J. N. Little and C. M. Thompson. *Control System Toolbox for Use with MATLAB. User's Guide.* MathWorks, 1992.

[13] J. M. Maciejowski. *Multivariable Feedback Design.* Addison-Wesley, Reading, MA, 1989.

[14] M. Morari and E. Zafiriou. *Robust Process Control.* Prentice Hall, Englewood Cliffs, NJ, 1989.

[15] K. Ogata. *Modern Control Engineering.* Prentice Hall, Englewood Cliffs, NJ, 1990.

[16] D. E. Seborg and T. F. Edgar and D. A. Mellichamp. *Process Dynamics and Control.* Wiley, New York, 1989.

[17] G. C. Stephanopoulous. *Chemical Process Control. An Introduction to Theory and Practice.* Prentice Hall, Englewood Cliffs, NJ, 1984.

3

Process identification

3.1 INTRODUCTION

This contribution is intended to give an appreciation of the scope and aims of identification, knowledge of some well-established identification techniques and some impression of the practical process of finding adequate model structures and acceptable parameter values. The coverage will be selective, partly because of constraints on space but more because the topic is not one which allows comprehensive formalization. One of the reasons why identification is such an interesting field is its diversity; new applications often require additions to existing techniques, and any classification of identification problems has to cover a large number of factors, as we shall see. On the other hand, the field is nearing maturity in that a range of approaches with relatively well understood strengths and weaknesses are available, and there is fairly broad agreement about them.

The intention is not to recommend "best" methods or examine current research, but to provide an acquaintance with the basis of identification, on which a more detailed knowledge can be built. The subject matter will be treated informally.

3.2 AIMS OF IDENTIFICATION

A process may be put to a wide range of possible uses: improving insight into system behaviour, prediction, control design, state estimation, simulation, fault diagnosis, etc. Each use conditions the form of the model and the criteria for its acceptance.

3.2.1 Insight

Mathematical modelling is the basis of "hard" science, where its role is to give insight into whatever is being investigated by providing a concise, quantified summary of the observed behaviour. A prior hypothesis about the causes may suggest a specific form for the model, reducing the identification procedure to estimation of parameter values. In such a case, accuracy of the parameter values may be the criterion for model quality.

Example: observations are made of the vertical position of a falling body. It is expected that its vertical acceleration is constant, so its position can be modelled as a quadratic in time, with acceleration, initial position and velocity as unknown (and constant) parameters.

Alternatively, the model may have a general structure intended to accommodate a wide range of possible behaviour and to be easy to analyse (a "black box" model).

Example: the body is not known to have a constant acceleration but is seen to move smoothly, so its position is modelled as a power series in time.

Conciseness is important in allowing insight. Indeed, a fundamentalist view is that a mathematical model is no more than a concise summary of the data. Unfortunately this does not address the question *how* concise the model should be, nor does it indicate how data of different types should be combined; for instance, prior (background or collateral) knowledge may well differ in form and reliability from the observations in a new experiment. It also ignores the fact that the form of a model is often dictated by the preferred physically motivated way of describing the system, and may well make gross simplifying assumptions.

Example: compartmental models are used in a number of biomedical applications to describe the dynamical behaviour of very complicated biological systems. For instance, the elimination of a drug by the liver might be modelled by a three-compartment model with one compartment each representing the blood, liver and gastro-intestinal system.

3.2.2 Prediction

Prediction is important in two ways for identification: as a criterion for validation of a model, and as the main object of developing a model. The use of prediction performance for validating a model is based on the idea that the model should describe the consistent, and hence predictable, part of the system's behaviour. Prediction performance is a sharper test than how well the model output fits the observed output. A too-large ("over-parameterized") model may be flexible enough to fit the observations well but may have some near-redundant combinations of parameters, determined mainly by inconsistent components of the observations such as noise or drift. These combinations have no predictive ability, yet will contribute to the predicted output, reducing its quality.

Predictive models also play an essential part in many control schemes. Even in classical frequency-response-based design, a Bode plot derived by sine-wave testing may be used to predict step response or noise sensitivity; the computation of the Bode plot relies on test techniques which minimize the effects of noise (e.g. use of a transfer-function analyser), implicitly separating consistent, predictable behaviour from inconsistent, unpredictable behaviour. Prediction is directly required in feed-forward control, predicting the later effect of a disturbance in order to cancel it by control action, and in many adaptive schemes, including self-tuning schemes such as generalized predictive control (GPC) and the large family of model predictive control schemes. In GPC and other schemes, an explicit separation of future output into predictable and unpredictable components is made, and the control action based on the predictable part.

3.2.3 State estimation

Recall that the state variables of a dynamical system are any set of variables of which the initial values are sufficient, together with the forcing, to determine the future state. A non-dynamical system is distinguished by not needing any such initial values, since its present output is completely determined by its present forcing. (Quite a lot of identification theory and technique does not depend on whether the system is dynamical or not; this is particularly true of the fundamentals of estimation theory). A huge variety of estimation problems in dynamical systems, including some parameter-estimation problems, can be posed as state estimation by regarding the unknowns as state variables. Examples include input estimation in environmental systems, deconvolution of distorted signals, target tracking, indirect or imperfect measurement as in biomass estimation in bioreactors, and condition monitoring and fault detection as in power systems and process plant. They all depend on a state-variable model, part or all of which may have to be identified rather than being known already.

As we shall see later, there is a close resemblance between state estimation and some parameter-estimation methods, and some technology can be shared by both. This is lucky, as a large amount of effort went into refining state-estimation techniques (mainly for aerospace applications) in the 1960s and 1970s; parameter estimation can borrow from this body of technique. There is one respect in which state estimation differs significantly from parameter estimation: the likelihood of ill-conditioned computation is quite high in state estimation, but low in parameter estimation so long as the model structure is well chosen. We shall look briefly at this and other differences later.

In both state estimation and identification, judgments have to be made about how much detail to put into the model, or more accurately how much detail can be left out. The presence of "process noise" in the dynamics equations of a state-variable model allows unmodelled forcing or ill-modelled dynamics such as mild non-linearities to be represented as additive stochastic terms, with specified (often guessed) means and covariances. Similarly "observation noise" may include modelling error as well as instrument noise, transmission or transcription errors and so on. Similar comments apply to identification, where "noise" may well cover systematic modelling errors, particularly when the model has to be kept simple for ease of analysis and design.

3.2.4 Simulation

This use of a model is not given much attention in the identification literature, yet it is one of the most widespread and important. For simulation, the model usually has to be in time-domain form, preferably as a set of difference and algebraic equations easily handled by a standard simulation package (although more flexible input facilities are now offered by some packages). The crucial aspect of many simulation models is their qualitative representativeness of the real system's behaviour, without necessarily being quantitatively accurate. However, some such models are used to explore "worst-case" scenarios, and must show the extremes of behaviour realistically. In others, average behaviour is to be examined, while the spread may also be of interest. We must ask whether the criterion for model performance can take such requirements into account. We shall discuss this further.

3.2.5 Fault detection

Faults in engineering systems may be detected by monitoring for unusual values of informative variables or by comparing various models representing faulted and unfaulted conditions. Both require reliable information on the ranges of normal and abnormal behaviour. In either case, the model should preferably employ parameters and/or variables with direct physical significance. This is a substantial constraint since, for instance, a standard and mathematically convenient discrete-time model structure, such as a rational Z-transform transfer function, may show the influence of a continuous-time parameter, say a time constant, in a complicated way involving several parameters. Another instance is that state variables chosen to give the state, forcing and observation matrices convenient structures may be more difficult to interpret than physically meaningful variables.

Other critical aspects of models for fault detection are their need to be updated on line rapidly enough to catch a fault promptly, ability to discriminate between a fault and a change in load (e.g. when a robot picks up an object), set point or input (e.g. feed quality in process plant), and reliability.

3.3 TYPES OF MODELS

The great majority of identification theory and technique is for a special family of dynamical models: linear, lumped, finite-order, discrete-time models. The main reason is that they are relatively easy to analyse and understand, and sufficiently versatile to be widely useful. We shall recall what "dynamical" means, then examine the properties assumed for this family of models, then look briefly at other properties and categories of model.

3.3.1 Dynamics

The output of a dynamical system at any instant depends on the system's history as well as its input at that time. It therefore has memory, often associated with stored energy or material. If only one independent variable (usually time) is considered, the system is lumped as distinct from distributed. If the present output can always be found from the input and a set of variables (its state) an infinitesimally short time ago, its state equation is a differential equation (generally vector) and its observation equation is just algebraic.

Example: The output voltage v(t) across the capacitor in a series RC circuit driven by a specified input e.m.f. waveform u(t) is $v(t) = \lim_{\delta t \to \infty}(v(t-\delta t) + \delta t(u(t-\delta t) - v(t-\delta t))/RC)$*, since in the limit* $\dot{v}(t) = (u(t) - v(t))/RC$*. The electric-field energy stored in the capacitor is* $Cv^2(t)/2$*, associated with state variable v(t). An initial value* $v(t_0)$ *and all subsequent forcing u(t),* $t \geq 0$*, is enough to determine the output for* $t \geq 0$*. The observation equation here is rather simple: output y(t) = v(t). Had the circuit contained only two resistors, it would not have been dynamical.*

The model order is the number of initial conditions needed to determine, with the forcing, the future output. Pure delay in the system causes the order to be infinite unless it is at the input or output and can be treated separately from finding the undelayed response; pure delays can, of course, be approximated by finite-order subsystems.

3.3.2 Linearity

As in control design, most identification theory and technique is for linear systems. The reasons are similar: difficulty in analysis, generalization and even categorization of non-linear systems, and the pre-existence of a powerful set of tools to deal with linear systems. A consequence of linearity is that we can identify a multi-input, multi-output (MIMO) system by adding single-input models together, since their effects can be separated then superposed, although this can lead to a larger number of states than required, particularly when the number of inputs and outputs grows. Equally important, in principle we can use any input signal which covers the system's bandwidth well enough, since the system model can be completely determined (in the absence of noise) from any such signal, and does not require specific amplitude characteristics. We therefore have considerable freedom to design a convenient signal with good excitation properties.

Example: valves in process plants virtually always have non-linear flow/pressure drop characteristics; but we may choose to treat the flow as linearly related to pressure drop for small changes about the normal regulated values.

3.3.3 Discrete- and continuous-time models

When discussing classical (pre-1965) identification methods, we shall be thinking about continuous-time models, but the great majority of computational identification algorithms assume that the model is discrete time, based on records with all variables sampled uniformly in time, at intervals of say T. Aliasing of the underlying continuous-time information will occur, scrambling the information in the signal, if the sampling is at a rate below the Nyquist rate (i.e. twice per cycle of the highest significant frequency present in the signal). Because of non-ideal filter and sampler characteristics, a more realistic rate is 10 samples per cycle. Account must be taken of, for instance, any resonance within the signal's roll-off range. Another factor to keep in mind is that a minimum-phase, continuous-time system may yield a non-minimum-phase, discrete-time model if sampled rapidly [1].

An important practical question is how to estimate continuous-time characteristics from discrete-time records. The converse problem, how to express prior continuous-time knowledge as constraints on a discrete-time model, is equally important. These questions have had surprisingly little attention in the identification literature. Use of a δ-operator model [4], in which the basic operator is $(z-1)/T$ rather than z, makes comparison with continuous-time behaviour easier, since the limit of δ as T tends to zero is the Laplace operator. The δ operator also has numerical advantages when the model includes a time constant much shorter than T, resulting in one or more poles very close to $z = 1$. Another possibility is to employ the bilinear operator $2(z-1)/(T(z+1))$. At frequencies considerably below $1/T$, the frequency response of a bilinear-operator model is close to that of the continuous-time model obtained by replacing the bilinear operator by s; consequently, we can safely interpret the discrete-time model as an approximate continuous-time model. This is less true of the δ-operator model, in which aliasing may introduce large discrepancies even at moderate frequencies.

We shall nevertheless mainly consider z-transform models, as most of the identification literature does. Often they will be rewritten as linear difference equations.

3.3.4 Time-invariance

Classical identification methods are exclusively for time-invariant models. Only in about the last 20 years have identification methods for time-varying models received much attention. The stimulus has been partly the use of models updated on line within adaptive control schemes; here we have to distinguish time variation due to the gradual reduction of initial errors and noise in the incoming observations from time variation due to representing the model parameters as varying, implying that observations become less relevant as they age.

It can be argued that no systems are time varying, since a sufficiently comprehensive model with constant coefficients should explain the observed behaviour. This is academic, of course, as it is often impracticable to identify such a model, but it does set time-varying models in context as models where time variation of the parameters is necessary to track behaviour which is otherwise omitted, such as non-linearity or high-order dynamics or the effects of unrecorded action elsewhere in the system. In fact, the use of linear, low-order, time-varying models to elucidate the behaviour of higher-order and/or non-linear systems is a fruitful idea, and we shall stress time-varying models for that reason.

3.3.5 Other properties of the model

We have several other choices to make about the type of model: should it be time or frequency domain, deterministic or stochastic, SISO or MIMO, input–output or state-variable, parametric or non-parametric? The best way to review these options is to see how they fit into the available identification methods and the application of the model.

Example: sine-wave testing is widely used to identify the gain and phase of a system at a number of frequencies. It yields an input–output system model (the frequency response) which is, in principle, non-parametric (a function of frequency, not a finite number of parameter values) and restricted to linear, time-invariant, lumped, SISO, deterministic systems. In practice, because the measurements are at a limited number of frequencies, the model is effectively parametric. It can, in fact, deal with quite noisy (stochastic) systems because the effects of noise can be much reduced by Fourier analysis of the output signal over a long enough period. Some non-linearity can also be tolerated because the harmonics generated by non-linearity do not contribute to the result so long as Fourier analysis is used to extract the fundamental from the response. MIMO systems can be handled simply by testing all combinations of input and output. Moreover, frequency-response models are popular because of the classical control-design methods based on them, and because people with an electrical engineering background find them familiar and easy to understand.

3.4 MODEL TYPES AND TEST SIGNALS

3.4.1 Non-parametric SISO time-domain models

Continuous-time dynamical systems adequately modelled by time-invariant linear models have total response equal to the sum of the forced response from zero initial conditions and the unforced response to the initial conditions. Most identification techniques assume that the response measured is the forced response, with a negligible contribution from the initial conditions; this assumption is clearly acceptable if the response recorded is much longer than the effective settling time of the system, or if the whole response excited by an input signal is recorded from an instant when the state of the system is quiescent (zero).

Recall that the forced (zero-i.c.) response of a linear, time-invariant, SISO, continuous-time system is

$$y(t) = \int_0^t h(t-\tau)u(\tau)d\tau \tag{3.1}$$

where u is the input, zero until time zero, h is the unit-impulse response and the state is zero at time zero. From (3.1) we see that we can find the output due to any known forcing if we know the impulse response. The impulse response is a function, specified in theory by its continuum of values over $-\infty \leq t \leq \infty$ and in practice by a large number of closely spaced values extending from time zero to the time when it has settled to near zero. It is thus a non-parametric description (although parameterized in practice). This has the advantage of not requiring (in theory) any model-structure selection (beyond a decision to assume linearity and time invariance). Its disadvantage is that for the response function to be specified well enough, a large number of values may have to be given, making a cumbersome description unsuitable for some purposes such as control design. Parametric models, discussed later, may be preferable.

Impulse testing

In practice, we have to add a noise term $v(t)$ to the right-hand side of (3.1) to account for instrument noise and error and modelling error such as ignored non-linearity or time-varying dynamics. The presence of this noise and, for some systems, inability to choose $u(t)$ freely are the only reasons we may have to devise identification methods cleverer than simply applying an impulse and measuring the response. The drawbacks of an impulse as an input for identification are: (1) its unrealisability, since a Dirac δ function has infinitesimal duration and infinite rate of change; an acceptable approximation is a pulse of finite rise, fall time and duration, all much less than the shortest feature of interest in $h(t)$; (2) the small energy available for a given peak size; we often cannot improve output signal:noise ratio (SNR) by increasing the size of a pulse input because that would lead to excitation of a non-linearity.

Step testing

The next simplest input is a step $u(t) \equiv \int_0^t \delta(\tau)d\tau$ giving a step response

$$s(t) = \int_0^t h(\tau)d(\tau) \tag{3.2}$$

since integration is a linear operation. In principle we need only differentiate $s(t)$ to find the impulse response. As a step is an infinite-energy, finite-power signal, a better SNR is obtainable than with an impulse input. However, since integration of a signal corresponds to multiplication of its Fourier spectrum by $1/jw$ and the spectrum of a δ function is flat, a step is deficient in high-frequency power, so the SNR of the high-frequency part of $y(t)$ may be poor. The result of this is to obtain a model which fits the lower-frequency dynamics better. We could consider averaging over a succession of steps making up a square wave; so long as the half-period were longer than the settling time of the system, the averaged step response would, without noise, be close to the actual step response, and the averaging would reduce the effects of noise which is inconsistent from step to step. This idea of averaging will appear again later.

Pulse testing

The discrete-time counterpart of (3.1) is

$$y_t = \sum_{i=0}^{t-1} h_{t-i} u_i \tag{3.3}$$

where the subscripts denote time in sampling intervals, sequence $\{h\}$ is the unit-pulse response and h_0 has been taken as zero, as no physical system responds instantly. Note that for a stable system, h_{t-i} is negligible for $t - i > t_s$ where t_s is the settling time of the system, so

$$y_t \simeq \sum_{i=t-t_s}^{t-1} h_{t-i} u_i \equiv \sum_{t'=1}^{t_s} h_{t'} u_{t-t'} \tag{3.4}$$

The practical usefulness of a pulse input in identification is limited in the same way as in the continuous-time case.

Sampled-step testing

The discrete-time step response, putting $u_i = 1, i \geq 0$ in (3.3), is

$$s_t = \sum_{i=0}^{t-1} h_{t-i} \tag{3.5}$$

and a sampled step input has similar advantages and disadvantages to its continuous-time version.

Correlation-based identification of the unit-pulse response

The need to minimize the effects of noise by ensuring a good SNR over the whole bandwidth of the system suggests we should try other input signals and/or other ways of processing the measurements. The basic idea of averaging out the effects of noise leads to a technique based on the relation between the unit-pulse response, the discrete-time input autocorrelation function $\{r_{uu,k}\} \equiv \{\varepsilon[u_t u_{t+k}]\}$ and the input–output cross-correlation function $\{r_{uy,k}\} \equiv \{\varepsilon[u_t y_{t+k}]\}$. (In our notation for these correlation functions, we have implicitly assumed that $\{\varepsilon[u_t u_{t+k}]\}$ and $\{\varepsilon[u_t y_{t+k}]\}$ are independent of time t; this property, together with time independence of the mean of $\{u\}$ and $\{y\}$, is called wide-sense stationarity.) If we shift (3.4) along by k samples, multiply by u_t and take expectations we get

$$r_{uy,k} = \varepsilon\left[\sum_{t'=1}^{t_s} h_{t'} u_t u_{t+k-t'}\right] = \sum_{t'=1}^{t_s} h_{t'} r_{uu,k-t'} \tag{3.6}$$

where expectation and summation have been interchanged, which is admissible since both are linear operations. As (3.6) is linear in the unknowns $h_{t'}$, $1 \leq t' \leq t_s$, we can solve it for $\{h\}$ by matrix inversion if $r_{uy,k}$ and the necessary $r_u y, k - t$'s are calculated for t_s values of k and if the r_{uu} values form a non-singular matrix.

The motive for employing (3.6) rather than (3.4) is that noise has less influence on (3.6); the additive noise v_{t+k} in the measured y_{t+k} is uncorrelated with the input, so it contributes nothing to $r_{uy,k} \equiv \varepsilon[u_t y_{t+k}]$ In practice, the expected values in the correlation functions are estimated as time averages

$$\hat{r}_{uu,k} = \frac{1}{N}\sum_{t=1}^{N} u_t u_{t+k} \simeq \varepsilon[u_t u_{t+k}]; \qquad \hat{r}_{uy,k} = \frac{1}{N}\sum_{t=1}^{N} u_t y_{t+k} \simeq \varepsilon[u_t y_{t+k}] \quad (3.7)$$

over a relatively long section of the $\{u\}$ and $\{y\}$ records, so the noise contributes finite but, we hope, small errors. Note that for (3.7) to be valid, $\{u\}$ and $\{y\}$ must be ergodic, i.e. have the same sample statistics when the sampling is over the same realization at various times as the statistics over the ensemble. Ergodicity is a stronger property than wide-sense (or, in fact, strict) stationarity.

Equation (3.6) is particularly convenient to use to identify $\{h\}$ when $\{r_{uu}\}$ is very simple. An extreme case is when $\{u\}$ is white noise, so that $r_{uu,k} = \sigma_u^2 \delta_k$, giving

$$r_{uy,k} = \sum_{t'=1}^{t_s} h_{t'} r_{uu,k-t'} = h_k \sigma_u^2 \quad (3.8)$$

The unit-pulse response estimates are then simply the input–output sample cross-correlation values. As white noise is not exactly realizable (having infinite bandwidth), a great deal of effort has gone into devising more convenient input signals. One important class of pseudo-noise (in the sense of having autocorrelation similar to white noise) signals is that of pseudo-random binary sequences (PRBSs).

Binary inputs are easy to generate and usually easy to apply to the system through the input actuator (although the actuator will impose its own dynamics on the signal). If a binary signal is zero-mean, it has maximal power per unit maximum amplitude, an important factor when the input is amplitude-constrained and the highest possible output SNR is required. On the other hand, the amplitude distribution is not typical of most practical inputs, which may mean that the effects of any non-linearity present are not typical. Moreover, the high rate of change at each binary transition will excite strongly any rate-dependent non-linearity, such as a limit on angular acceleration in a drive system, resulting from a limit on motor armature current and/or magnetic field strength and hence torque.

Among the most popular PRBSs are random telegraph waves and maximal-length sequences (m-sequences). In random telegraph waves the binary state changes at apparently random instants at intervals with a specified probability distribution. In an m-sequence the switching is synchronous; the binary state is constant over a fixed bit interval. The sequence of bit values has a period of $2^N - 1$ bits, and the $2^N - 1$ successive N-bit words starting within one period include every possible N-bit word except N zeros. The autocorrelation function of an m-sequence with binary levels $\pm A$ is A^2 at lags $k = m(2^N - 1)$ where m is any integer, and $-A^2/(2^N - 1)$ at all other lags. Clearly $\{h\}$ is readily computed from $\{r_{uy}\}$ computed over a whole number of periods, so long as the settling time t_s is less than the period of the m-sequence, since $\{r_{uy}\}$ is then a constant plus a scale factor times $\{h\}$. The constant may be avoided by using asymmetrical binary levels.

If $2^N - 1$ is large enough, $-a^2/(2^N - 1)$ is negligible and $\{h\}$ is proportional to $\{r_{uy}\}$ approximately. In other words, the cross-correlation, and therefore the output, is approximately that which would result from a white-noise input, implying that the power spectrum of the m-sequence is approximately flat over the bandwidth of the system. This can be verified theoretically (by using the Wiener–Khinchine relation), but it is important to remember that the flatness of the power spectrum applies only to a whole number of periods; there is no guarantee that a non-integral number of periods of an m-sequence has good spectral or autocorrelation properties. The bit interval is determined by the time resolution required and the number of samples (bits) per period is constrained to 1, 3, 7, 15, 31, 63, 127,..., so the experimental conditions may seem rather restrictive. However, the experiment must be long enough for the initial-condition response to have died to much less than the forced response, so the minimum experiment length does not depend only on the sequence parameters. A further practical point is that some non-linearities have effects on m-sequence results which may be difficult to distinguish, without other evidence, from more complicated linear dynamics.

3.4.2 Non-parametric SISO frequency-domain models

If we take the Fourier transform of (3.1)

$$y(t) = \int_0^t h(t-\tau)u(\tau)d\tau \tag{3.9}$$

and assume zero initial conditions, we get

$$Y(j\omega) = H(j\omega)U(j\omega) \tag{3.10}$$

If the input $u(t)$ is a sinusoid, which has a complex spike (δ-function) positive-frequency spectrum, the output spectrum is a scaled and generally phase-shifted spike at the same frequency, and we have

$$\mid H(j\omega) \mid = \frac{\mid Y(j\omega) \mid}{\mid U(j\omega) \mid}, \qquad \angle H(j\omega) = \angle Y(j\omega) - \angle U(j\omega) \tag{3.11}$$

Thus we can identify points on the complex Fourier transfer function $H(j\omega)$ by applying sinusoids at successive frequencies, waiting each time for the initial-condition response to die away, and noting the amplitude ratio and phase shift across the system. All of this is elementary to electrical and control engineers, and the Fourier transfer function is the basis of classical control design. Nevertheless, we shall regard it as quite a sophisticated identification technique with several advantages; it has several significant drawbacks, too.

Let us look first at the advantages of sine-wave testing, particularly as implemented by transfer-function analysers. It is worth noting that direct measurement of gain and phase change is sensitive to modest amounts of drift, bias and distortion. Transfer-function analysers do not measure the gain $\mid Y(j\omega) \mid / \mid U(j\omega) \mid$ and phase change directly but instead carry out the operation of Fourier analysis. This consists of multiplying the output separately by the input and a signal 90^0 ahead of the input in phase and integrating each product waveform over an integral number of periods, yielding the "sine" and "cosine" coefficients of the fundamental term of the Fourier series of the output. The main reason for doing this more complicated processing is that the integration, a low-pass filtering operation, attenuates the effects of wideband noise. Integrating the product of a reference sinusoid and the output over an whole number of periods also confers other benefits: any output component of the output at a harmonic frequency of the input, such as would be generated by a non-linearity, gives zero result, and so does any constant component in the output. A discrete-time transfer-function analyser is easily realized by software, employing a sampling frequency much higher than the highest input frequency used.

Sine-wave testing is still very widely used (for instance, as "wobble tests" in the aero-engine industry), but its disadvantages counterbalance the advantages just mentioned. The most obvious is the time taken to carry out a range of single-frequency tests large enough to cover the system's bandwidth at high enough resolution in frequency, particularly for slow systems with time constants of hours or days. Another is the difficulty or impossibility of producing a reasonably accurate sine wave with many input actuators; for instance, many valve controllers in process plant only allow fixed-rate opening and closing. A third is the possibility that important behaviour such as a resonance will be missed by unlucky choice of frequency spacing. A fourth is that dead time is identified only by its effect on phase, and may be difficult to estimate well in the face of phase-measurement inaccuracy, particularly if the gain roll-off rate is high. Finally, the result of sine-wave testing is a frequency-sampled frequency–response function, an uneconomical description consisting of many numbers giving limited insight and cumbersome as a basis for subsequent use such as prediction or on-line control synthesis.

3.4.3 Parametric SISO time-domain models

The unit-impulse or unit-pulse response may be approximated by the sum of components such as exponentials:

$$h(t) = \sum_{i=1}^{n} f_i \exp\left(g_i(t - t_{di})\right) \mu(t - t_{di}) \tag{3.12}$$

where the dead times t_{di} may be negligible and often all t_{di} are the same, the g_i are complex in general but often real and negative and similarly for f_i. This parametric model requires the selection of a structural parameter n. In discrete time, (3.12) has the counterpart

$$h_t = \sum_{i=1}^{n} \beta_i \gamma_i^{t-d_i} \mu_{t-d_i} \tag{3.13}$$

where μ_0 is the sampled unit step function. In these forms the models are usable but not very convenient to handle, so their Laplace- or z-transform versions are more often employed. One of the main reasons why (3.12) and (3.13) are not convenient is their non-linearity in all parameters except f_i and β_i. Linearity in the parameters makes parameter estimation much easier, as we shall see later. Their transform versions are still partly non-linear but turn out to be easy to rewrite in linear form.

3.4.4 SISO transform models

Laplace transforming (3.12)

$$H(s) = \sum_{i=1}^{n} \frac{f_i}{s - g_i} \exp(-s t_{di}) \tag{3.14}$$

and if all the dead times are t_d,

$$H(s) = \frac{\sum_{i=1}^{n} f_i \prod_{k=1,k\neq i}^{n}(s - g_k)}{\prod_{i=1}^{n}(s - g_i)} \exp(-st_d) \tag{3.15}$$

Apart from the exponential, this is a rational polynomial function of s. The input–output relation (omitting noise) is

$$Y(s) = H(s)U(s) \tag{3.16}$$

and multiplying by the denominator of $H(s)$ gives

$$\prod_{i=1}^{n}(s - g_i)Y(s) = \left(\sum_{i=1}^{n} f_i \prod_{k=1,k\neq i}^{n} (s - g_k)\exp(-st_d)\right) U(s) \tag{3.17}$$

This is in the form

$$A(s)Y(s) = B(s)\exp(-st_d)U(s) \tag{3.18}$$

where $A(s)$ and $B(s)$ are polynomials of degree n and $n-1$ in s. Defining some new notation and renormalizing turns it into

$$\left(1 + \sum_{i=1}^{n} p_i s^i\right) y(s) = \left(\sum_{i=0}^{n-1} q_i s^i \exp(-st_d)\right) U(s) \tag{3.19}$$

which is linear in all parameters except t_d. The inverse transform of (3.19) is

$$1 + \sum_{i=1}^{n} p_i y^{(i)}(t) = \sum_{i=0}^{n-1} q_i u^{(i)}(t - t_d) \tag{3.20}$$

from which it can be seen that the pure delay t_d can readily be taken out of the model by delaying the input record by t_d. The only thing then preventing us from identifying the model easily from (3.20), for instance by linear least-squares estimation, is the practical difficulty, in the presence of wideband noise, of producing good measurements of the derivatives $u^{(i)}(t - t_d)$ and $y^{(i)}(t)$. Sometimes we can do it well enough by using "state-variable filters" which approximately differentiate the signals over a limited bandwidth, so that the amplification of the noise is not unacceptably high. Another clear disadvantage of models in the form (3.20) is that the physically meaningful parameters $\{f_i\}$ and $\{g_i\}$ have been combined into p_i and q_i, which are not easily interpreted. Once $\{p_i\}$ and $\{q_i\}$ have been estimated, $\{f_i\}$ and $\{g_i\}$ can be recovered in a simple calculation, but because of the non-linearity of the relation between the two sets of parameters, good estimates of $\{p_i\}$ and $\{q_i\}$ do not necessarily convert into good estimates of $\{f_i\}$ and $\{g_i\}$.

An exactly similar derivation in discrete time, using z transforms instead of Laplace transforms, gives the linear-in-parameters z-transform model

$$\begin{aligned}\left(1 + \sum_{i=1}^{n} a_i z^{-i}\right) Y(z^{-1}) &= \left(\sum_{i=0}^{n-1} b_i z^{-i}\right) z^{-d}U(z^{-1}) \\ &\equiv A(z^{-1})Y(z^{-1}) = B(z^{-1})z^{-d}U(z^{-1})\end{aligned} \tag{3.21}$$

where the dead time is d sample intervals. Inverting the z transforms or interpreting z^{-1} as a one-sample-delay operator (which should strictly be distinguished from $z = \exp(-sT)$ and is usually denoted by q), (3.21) gives

$$y_t + \sum_{i=1}^{n} a_i y_{t-i} = \sum_{i=0}^{n-1} b_i u_{t-i-d} \tag{3.22}$$

As no physical system has instantaneous response, b_0 is zero and d denotes the dead time beyond the minimum. We can rearrange (3.22) as a linear regression equation

$$y_t = -\sum_{i=1}^{n} a_i y_{t-i} + \sum_{i=1}^{n} b_i u_{t-d-i+1} \equiv \boldsymbol{\Phi}_t^T \boldsymbol{\Theta} \tag{3.23}$$

with

$$\begin{aligned} \boldsymbol{\phi}_t^T &\equiv [(-y_{t-1})\,(-y_{t-2})\,\ldots\,(-y_{t-n})\; u_{t-d}\; u_{t-d-1}\,\ldots\, u_{t-d-n+1}], \\ \boldsymbol{\theta}^T &\equiv [a_1\, a_2 \ldots\, a_n\; b_1\, b_2 \ldots\, b_n] \end{aligned} \tag{3.24}$$

Such a model is in the right form for least-squares estimation of the parameters making up θ, but (3.23) does not include any noise term. There are several possible ways to add noise to a linear model like (3.23). For the moment, let us not worry about the exact nature of the noise but just represent it as an additive term v_t:

$$y_t \equiv \boldsymbol{\phi}_t^T \boldsymbol{\theta} + v_t \tag{3.25}$$

This is called an equation-error model (also ARX-model), as distinct from an output-error model, also based on (3.21),

$$Y(z^{-1}) = \frac{B(z^{-1})z^{-d}}{A(z^{-1})} U(z^{-1}) + V_o(z^{-1}) \tag{3.26}$$

The main significance of the difference is that in (3.26) the noise $\{v_o\}$ is represented as not having passed through any of the system's input–output dynamics. On the other hand, if we take (3.21) and add noise as in (3.25), we get

$$A(z^{-1})Y(z^{-1}) = B(z^{-1})z^{-d}U(z^{-1}) + V(z^{-1}) \tag{3.27}$$

then dividing by $A(z^{-1})$ gives

$$Y(z^{-1}) = \frac{B(z^{-1})z^{-d}}{A(z^{-1})} U(z^{-1}) + \frac{1}{A(z^{-1})} V(z^{-1}) \tag{3.28}$$

which causes problems in least-squares estimation. The output-error model (3.26) is less widely used than the equation-error model (3.27) because the latter, written in the form (3.25), is linear in its parameters, making parameter estimation easier, while (3.26) is non-linear in the coefficients in $A(z^{-1})$.

In Section 3.6 we will look at computational algorithms to estimate $\boldsymbol{\theta}$ in (3.25).

3.4.5 MIMO models

We shall not look explicitly at multivariable models, although most of the estimation theory extends easily to them and the parameter-estimation algorithms deal readily with multi-input, single-output (MISO) systems. The only reason why a multivariable system may have to be treated as such rather than as a number of MISO systems for identification is that the model has to be *parsimonious*, i.e. economical in the number of parameters (either for convenience in use or to achieve high statistical efficiency). For example, a pole shared between two or more input–output relations appears only once in a state-variable model but as many times as the number of outputs involved if MISO models are used.

The identification literature contains only a relatively small amount on MIMO models, partly because, as just implied, they can often be avoided by MISO or even SISO modelling, partly because selection of model form and parameterization is not a trivial matter for multivariable systems, and partly because some parameter-estimation algorithms which are convenient and effective for SISO or MISO models are unable to handle MIMO models. A further practical factor is that development of a model is an interactive process, with the modeller having to make decisions founded on an understanding of the system and the model's resemblances to and deviations from it; this is much harder to achieve for MIMO systems and models than for SISO or MISO ones. A second practical reason for confining models to modest numbers of variables is that data-logging for a smaller number is less likely to be affected by instrument or system malfunctions or breakdowns.

Next we examine some popular identification algorithms.

3.5 IDENTIFICATION ALGORITHMS

3.5.1 Batch ordinary least-squares estimation

Least-squares estimation finds the value of vector $\boldsymbol{\theta}$ in the linear-in-parameters model

$$y_t = \boldsymbol{\phi}_t^T \boldsymbol{\theta}_t + e_t \qquad t = 1, 2, \ldots, N \tag{3.29}$$

which minimizes the sum of the squares of the model-output errors $\{e_t\}$. Writing model (3.29) as

$$\boldsymbol{y} = \boldsymbol{\Phi}\boldsymbol{\theta} + \boldsymbol{e} \tag{3.30}$$

the sum of squared errors is

$$S_N = \boldsymbol{e}^T \boldsymbol{e} \tag{3.31}$$

The minimum is

$$\frac{\partial S_N}{\partial \boldsymbol{\theta}} = \frac{\partial}{\partial \boldsymbol{\theta}} \left\{ (\boldsymbol{y} - \boldsymbol{\Phi}\boldsymbol{\theta})^T (\boldsymbol{y} - \boldsymbol{\Phi}\boldsymbol{\theta}) \right\} = 2(-\boldsymbol{\Phi}^T \boldsymbol{y} + \boldsymbol{\Phi}^T \boldsymbol{\Phi}\boldsymbol{\theta}) = 0 \tag{3.32}$$

so the minimizing estimate $\hat{\boldsymbol{\theta}}$ is

$$\hat{\boldsymbol{\theta}} = [\boldsymbol{\Phi}^T \boldsymbol{\Phi}]^{-1} \boldsymbol{\Phi}^T \boldsymbol{y} \tag{3.33}$$

Notice that the dimensions of the *normal matrix* $\boldsymbol{\Phi}^T\boldsymbol{\Phi}$ to be inverted in (3.33) are equal to the number of elements in $\boldsymbol{\theta}$, usually quite small, and not dependent on the number of observations N, usually large. However, the computational labour in forming the matrix is large if N is large. Note also that the estimate is linear in the observations making up $\boldsymbol{y}$. The explanatory variables making up $\boldsymbol{\Phi}$ are called *regressors* and enter (3.33) non-linearly; they include observations of the system output if the model is autoregressive. Least-squares estimation is still sometimes called *regression*, a name from its early use in statistical genetic studies.

The linear equations for $\boldsymbol{\theta}$ in (3.32),

$$\boldsymbol{\Phi}^T\boldsymbol{\Phi}\boldsymbol{\theta} = \boldsymbol{\Phi}^T\boldsymbol{y} \tag{3.34}$$

are the *normal equations*. If the regressors are strongly interrelated, the equations are ill conditioned and the inversion of the normal matrix may give unacceptable sensitivity to small errors. There are several ways to reduce or avoid such difficulties. One is to factorize the normal matrix and carry out the inversion with the factors; if, for example, matrix square roots such as Cholesky factors are employed, their condition number is the square root of that of the normal matrix, reducing ill conditioning accordingly. Alternatively, the singular-value decomposition of the normal matrix can be used to allow computation of $\hat{\boldsymbol{\theta}}$ without explicit inversion of the normal matrix. This technique has the further advantage of showing, through the smallest singular values, which combinations of regressors have little explanatory power and should be removed from the model to improve conditioning; removal amounts to merely setting small singular values to zero. The view to take of ill conditioning in least-squares estimation is as a valuable indication of a weakness (specifically a near-redundancy) in the model structure, rather than as just a numerical difficulty to be cured.

The basic formula (3.33) is called *ordinary least squares (OLS)* to distinguish it from more elaborate versions of least-squares estimation.

An important property of the OLS estimate is the orthogonality of the model-output estimate to the model-output error:

$$(\boldsymbol{\Phi}\hat{\boldsymbol{\theta}})^T\,(\boldsymbol{y} - \boldsymbol{\Phi}\hat{\boldsymbol{\theta}}) = \hat{\boldsymbol{\theta}}^T(\boldsymbol{\Phi}^T\boldsymbol{y} - \boldsymbol{\Phi}^T\boldsymbol{\Phi}\hat{\boldsymbol{\theta}}) = 0 \tag{3.35}$$

Furthermore, we can easily show that the error is orthogonal to each regressor.

The OLS estimate is unbiased if the error is zero-mean and independent of the regressors, since then

$$\begin{aligned}\varepsilon[\hat{\boldsymbol{\theta}}] &= \varepsilon[[\boldsymbol{\Phi}^T\boldsymbol{\Phi}]^{-1}\boldsymbol{\Phi}^T(\boldsymbol{\Phi}\boldsymbol{\theta} + \boldsymbol{e})] = \varepsilon[\boldsymbol{\theta}] + \varepsilon[[\boldsymbol{\Phi}^T\boldsymbol{\Phi}]^{-1}\boldsymbol{\Phi}^T\boldsymbol{e}] \\ &= \varepsilon[\boldsymbol{\theta}] + \varepsilon[[\boldsymbol{\Phi}^T\boldsymbol{\Phi}]^{-1}\boldsymbol{\Phi}^T]\varepsilon[e] = \varepsilon[\boldsymbol{\theta}]\end{aligned} \tag{3.36}$$

In fact, we can slightly weaken the assumption that $\{e\}$ is independent of the regressors by considering probability limits; it is fairly obvious that $plim\ \hat{\boldsymbol{\theta}} = plim\ \boldsymbol{\theta}$ so long as $plim\ \boldsymbol{\Phi}^T\boldsymbol{e}$ is zero.

Another helpful feature of OLS is the easy availability of an estimate of the covariance of the estimate. If the noise $\{e_t\}$ is zero-mean, white (serially uncorrelated) and has constant variance σ^2, and $\boldsymbol{\Phi}$ is regarded as deterministic, then $cov\ \boldsymbol{e} = \sigma^2\boldsymbol{I}$ and

$$\begin{aligned}cov\ \hat{\boldsymbol{\theta}} &= \varepsilon[(\hat{\boldsymbol{\theta}} - \boldsymbol{\theta})(\hat{\boldsymbol{\theta}} - \boldsymbol{\theta})^T] = \varepsilon[([\boldsymbol{\Phi}^T\boldsymbol{\Phi}]^{-1}\boldsymbol{\Phi}^T\boldsymbol{y} - \boldsymbol{\theta})([\boldsymbol{\Phi}^T\boldsymbol{\Phi}]^{-1}\boldsymbol{\Phi}^T\boldsymbol{y} - \boldsymbol{\theta})^T] \\ &= \varepsilon[([\boldsymbol{\Phi}^T\boldsymbol{\Phi}]^{-1}\boldsymbol{\Phi}^T(\boldsymbol{\Phi}\boldsymbol{\theta} + \boldsymbol{e}) - \boldsymbol{\theta})([\boldsymbol{\Phi}^T\boldsymbol{\Phi}]^{-1}\boldsymbol{\Phi}^T(\boldsymbol{\Phi}\boldsymbol{\theta} + \boldsymbol{e}) - \boldsymbol{\theta})^T] \\ &= \varepsilon[([\boldsymbol{\Phi}^T\boldsymbol{\Phi}]^{-1}\boldsymbol{\Phi}^T\boldsymbol{e})([\boldsymbol{\Phi}^T\boldsymbol{\Phi}]^{-1}\boldsymbol{\Phi}^T\boldsymbol{e})^T]\end{aligned}$$

$$= \left[\boldsymbol{\Phi}^T\boldsymbol{\Phi}\right]^{-1}\boldsymbol{\Phi}^T\varepsilon[ee^T]\boldsymbol{\Phi}[\boldsymbol{\Phi}^T\boldsymbol{\Phi}]^{-1} = \sigma^2[\boldsymbol{\Phi}^T\boldsymbol{\Phi}]^{-1} \tag{3.37}$$

so the only extra labour involved in finding $cov\ \hat{\boldsymbol{\theta}}$ is the provision of an estimate of the noise variance σ^2 and, if explicit inversion of the normal matrix has been avoided, the recovery of the inverse.

3.5.2 Batch weighted-least-squares estimation

The sum-of-squares cost function S_N can be generalized by attaching a non-negative weight w_t to e_t^2 giving

$$S_N(\boldsymbol{W}) = \boldsymbol{e}^T\boldsymbol{W}\boldsymbol{e} \qquad \text{with} \qquad \boldsymbol{W} = diag(w_t) \tag{3.38}$$

Algebra similar to that for OLS gives the *weighted least-squares (WLS)* estimate

$$\hat{\theta}(\boldsymbol{W}) = [\boldsymbol{\Phi}^T\boldsymbol{W}\boldsymbol{\Phi}]^{-1}\boldsymbol{\Phi}^T\boldsymbol{W}\boldsymbol{y} \tag{3.39}$$

An obvious application of WLS estimation is to take account of differences in the reliability among the observations, for instance due to varying noise levels; a larger weight is attached to the error for a more reliable observation. This idea will appear again when we consider minimum-covariance estimation.

3.5.3 Recursive ordinary-least-squares estimation

The estimation algorithms presented so far have computed the estimate from all the data in one go. For some purposes it is more convenient, or even essential, to do the estimation *recursively*, i.e. taking the observations one at a time (usually in chronological order) and updating the estimate and perhaps also its estimated covariance each time. This allows on-line estimation, for instance as part of an adaptive control scheme, but on-line estimation requires a number of safeguards to avoid inappropriate and potentially disastrous updating gains and is less straightforward than it appears at first sight. A second, arguably more important, use for recursive estimation is to permit a time-varying model to track behaviour which is too complicated to be described adequately by a constant-parameter model; for instance, non-linear systems can often be modelled in this way, and the nature of the non-linearity may be clear from the time variation of the temporarily valid linear model. An example is a seasonally or diurnally varying environmental model.

We shall consider the scalar-output linear model (3.29) again, and assume that an initial estimate $\hat{\theta}_0$ is available at time zero. It will probably not be accurate, and its uncertainty is indicated by an initial covariance $\boldsymbol{P}_0$ if the uncertainty is itself uncertain, $\boldsymbol{P}_0$ is simply set to a very large value, say $10^6\boldsymbol{I}$, with the effect that after a few observations have been processed, the influence of the initial estimate will be small.

Consider the changes when a new observation y_t and a new regressor vector ϕ_t are to update the OLS estimate (3.33)

$$\hat{\theta}_{t-1} = [\boldsymbol{\Phi}_{t-1}^T\boldsymbol{\Phi}_{t-1}]^{-1}\boldsymbol{\Phi}_{t-1}^T y_{t-1} \tag{3.40}$$

previously computed after receiving y_{t-1} and ϕ_{t-1}. Denoting the estimated covariance $\sigma^2[\boldsymbol{\Phi}_{t-1}^T\boldsymbol{\Phi}_{t-1}]^{-1}$ by $\boldsymbol{P}_{t-1}$ and noting that

$$\boldsymbol{\Phi}_{t-1}^T\boldsymbol{\Phi}_{t-1} = \sum_{i=1}^{t-1}\phi_i\phi_i^T \qquad \text{and} \qquad \boldsymbol{\Phi}_{t-1}^T y_{t-1} = \sum_{i=1}^{t-1}\phi_i y_i \tag{3.41}$$

we have

$$\hat{\boldsymbol{\theta}}_t = \frac{\boldsymbol{P}_t}{\sigma^2}\left(\sum_{i=1}^{t-1}\phi_i y_i + \phi_t y_t\right) = \frac{\boldsymbol{P}_t}{\sigma^2}\left(\sum_{i=1}^{t-1}\phi_i\phi_i^T\hat{\boldsymbol{\theta}}_{t-1} + \phi_t y_t\right)$$

$$= \frac{\boldsymbol{P}_t}{\sigma^2}\left((\sigma^2\boldsymbol{P}_t^{-1} - \phi_t\phi_t^T)\hat{\boldsymbol{\theta}}_{t-1} + \phi_t y_t\right) = \hat{\boldsymbol{\theta}}_{t-1} + \frac{\boldsymbol{P}_t}{\sigma^2}(y_t - \phi_t^T\hat{\boldsymbol{\theta}}_{t-1}) \tag{3.42}$$

The new estimate of $\boldsymbol{\theta}$ is thus the old estimate plus a correction proportional to the output-prediction error given by the old estimate. Furthermore, the correction gain is proportional to the uncertainty in the new estimate and inversely proportional to the noise variance. All these features seem intuitively reasonable (although we might have guessed that the correction gain should be proportional to the covariance of the old estimate).

We also need an updating equation for $\boldsymbol{P}$. Write $\boldsymbol{P}_{t-1}$ as $\boldsymbol{P}_{t-1}\boldsymbol{P}_t^{-1}\boldsymbol{P}_t$, then

$$\boldsymbol{P}_{t-1} = \boldsymbol{P}_{t-1}(\boldsymbol{P}_{t-1}^{-1} + \frac{\phi_t\phi_t^T}{\sigma^2})\boldsymbol{P}_t = \boldsymbol{P}_t + \frac{\boldsymbol{P}_{t-1}\phi_t\phi_t^T\boldsymbol{P}_t}{\sigma^2} \tag{3.43}$$

so

$$\phi_t^T\boldsymbol{P}_{t-1} = \phi_t^T\boldsymbol{P}_t + \frac{\phi_t^T P_{t-1}\phi_t\phi_t^T\boldsymbol{P}_t}{\sigma^2} \tag{3.44}$$

giving

$$\phi_t^T\boldsymbol{P}_t = \frac{\phi_t^T\boldsymbol{P}_{t-1}}{1 + \frac{\phi_t^T\boldsymbol{P}_{t-1}\phi_t}{\sigma^2}} \tag{3.45}$$

Substitution of (3.45) into the last term of (3.43) then yields

$$\boldsymbol{P}_t = P_{t-1} - \frac{\boldsymbol{P}_{t-1}\phi_t\phi_t^T\boldsymbol{P}_{t-1}}{\sigma^2} = \boldsymbol{P}_{t-1} - \frac{\boldsymbol{P}_{t-1}\phi_t\phi_t^T\boldsymbol{P}_{t-1}}{\sigma^2 + \phi_t^T\boldsymbol{P}_{t-1}\phi_t} \tag{3.46}$$

Equations (3.42) and (3.46) allow observation-by-observation updating of the estimates and their estimated covariance. They are a little easier to assimilate if written as

$$\begin{aligned}
\hat{y}_t &= \phi_t^T\hat{\boldsymbol{\theta}}_{t-1} \\
v_t &= y_t - \hat{y}_t \\
f_t &= \phi_t^T\boldsymbol{P}_{t-1}\phi_t \\
\boldsymbol{P}_t &= \boldsymbol{P}_{t-1} - \frac{\boldsymbol{P}_{t-1}\phi_t\phi_t^T\boldsymbol{P}_{t-1}}{1 + f_t} \\
\boldsymbol{g}_t &= \boldsymbol{P}_t\phi_t \\
\hat{\boldsymbol{\theta}}_t &= \hat{\boldsymbol{\theta}}_{t-1} + \boldsymbol{g}_t v_t
\end{aligned}$$

The computational load of this algorithm is small; no matrix inversion arises, and once $\boldsymbol{P}_{t-1}\phi_t$ has been computed, only vector and scalar operations are needed.

3.5.4 Recursive weighted-least-squares estimation

Weight w_t may be attached to the error at time t by modifying the algorithm to

$$\begin{aligned}
\hat{y}_t &= \boldsymbol{\phi}_t^T \hat{\boldsymbol{\theta}}_{t-1} \\
v_t &= y_t - \hat{y}_t \\
f_t &= w_t \boldsymbol{\phi}_t^T \boldsymbol{P}_{t-1} \boldsymbol{\phi}_t \\
\boldsymbol{P}_t &= \boldsymbol{P}_{t-1} - \frac{w_t \boldsymbol{P}_{t-1} \boldsymbol{\phi}_t \boldsymbol{\phi}_t^T \boldsymbol{P}_{t-1}}{1 + f_t} \\
\boldsymbol{g}_t &= w_t \boldsymbol{P}_t \boldsymbol{\phi}_t \\
\hat{\boldsymbol{\theta}}_t &= \hat{\boldsymbol{\theta}}_{t-1} + \boldsymbol{g}_t v_t
\end{aligned}$$

This algorithm is useful for tracking time-varying systems (see Section 3.5.7) and for achieving minimum covariance of parameter error when the observation noise is white.

3.5.5 Information updating

The inverse $\boldsymbol{P}^{-1}$ of the parameter-error covariance is called the *information matrix* and for Gaussian $\hat{\boldsymbol{\theta}}$ is in fact the Fisher information matrix. The update from $\boldsymbol{P}_{t-1}$ to $\boldsymbol{P}_t$ is related by the famous matrix-inversion lemma $(\boldsymbol{A}+\boldsymbol{B}\boldsymbol{C})^{-1} = \boldsymbol{A}^{-1} - \boldsymbol{A}^{-1}\boldsymbol{B}(\boldsymbol{I}+\boldsymbol{C}\boldsymbol{A}^{-1}\boldsymbol{B})\boldsymbol{C}\boldsymbol{A}^{-1}$ to the information-updating equation

$$\boldsymbol{P}_t^{-1} = \boldsymbol{P}_{t-1}^{-1} + w_t \boldsymbol{\phi}_t \boldsymbol{\phi}_t^T \tag{3.47}$$

This is readily interpreted as adding the information conveyed by the observation at time t, duly weighted, to the previous information. If we choose as w_t the inverse $1/\sigma_t^2$ of the observation-noise variance, the new information is proportional to the "mean-square signal:noise ratio" of the new observation, since $\boldsymbol{\phi}_t \boldsymbol{\phi}_t^T$ gives the "mean-square" clean output (normalized by $\boldsymbol{\theta}$).

Recursive least squares may be implemented with information updating in place of covariance updating. This is commonly done in its state-estimation counterpart, but not usually in parameter estimation. In state estimation, numerical conditioning problems can arise in covariance updating because of almost noise-free observations introducing near-singularity into the covariance matrix. In parameter estimation, by contrast, the model is virtually always underparameterized, in the sense that the model-output error includes a systematic component due to neglected behaviour such as non-linearity, higher-order dynamics or the effects of unmonitored inputs. There is thus little risk of observations being almost noise-free.

3.5.6 Algorithms to avoid bias due to correlation between regressors and noise

The problem

As seen in Section 3.5.1, the OLS estimate is unbiased if the noise is zero-mean and independent of the regressors. A reasonable test of a good model, in theory, is whether it leaves the model-output errors (or *residuals*, which can be thought of as estimates of the noise) not related to the regressors (explanatory variables); any systematic relation would indicate failure of the model to explain all the influence of the regressors on the output. In practice, things are not so clear-cut, as the model structure may be too restrictive to allow all the influence to be represented. An obvious example is where the dynamical order has been restricted to make the model easier to use, omitting some non-dominant dynamics. Another is the presence of non-linearity, making it impossible to explain all the effect of the regressors on the output with a linear model. We should therefore expect some correlation between regressors and error in many cases (although if the correlation is at a lag not equal to any regressor-to-error lag in the model, $\varepsilon[\boldsymbol{\Phi}]$ is zero and bias is not implied). Correlation between regressors and noise is introduced if a rational transfer-function model is rewritten as a linear difference equation in the form (3.29). Consider the output-error model

$$Y(z^{-1}) = \frac{B(z^{-1})z^{-d}}{A(z^{-1})}U(z^{-1}) + V_o(z^{-1}) \tag{3.48}$$

rewritten in equation-error form as

$$\begin{aligned} A(z^{-1})Y(z^{-1}) &= B(z^{-1})z^{-d}U(z^{-1}) + A(z^{-1})V_o(z^{-1}) \\ &\equiv B(z^{-1})z^{-d}U(z^{-1}) + E(z^{-1}) \end{aligned} \tag{3.49}$$

and hence as

$$y_t = \boldsymbol{\phi}_t^T \boldsymbol{\theta}_t + e_t \qquad t = 1, 2, \ldots, N \tag{3.50}$$

with

$$\begin{aligned} \boldsymbol{\phi}_t^T &\equiv [(-y_{t-1})\ (-y_{t-2})\ \ldots\ (-y_{t-n})\ u_{t-d}\ u_{t-d-1}\ \ldots\ u_{t-d-n+1}], \\ \boldsymbol{\theta}^T &\equiv [a_1\ a_2 \ldots\ a_n\ b_1\ b_2 \ldots\ b_n] \end{aligned}$$

Here the noise sequence $\{e\}$ is autocorrelated even if $\{v_o\}$ was white, and that is enough to imply correlation with $\{\boldsymbol{\phi}\}$, since for instance e_{t-1} is part of y_{t-1} in $\boldsymbol{\phi}_t$ and is correlated with e_t. Because this implies correlation between $\boldsymbol{e}$ and $\boldsymbol{\Phi}$, we can no longer say that

$$\begin{aligned} \varepsilon[\hat{\boldsymbol{\theta}}] &= \varepsilon\left[[\boldsymbol{\Phi}^T\boldsymbol{\Phi}]^{-1}\boldsymbol{\Phi}^T(\boldsymbol{\Phi}\boldsymbol{\theta} + \boldsymbol{e})\right] = \varepsilon[\boldsymbol{\theta}] + \varepsilon\left[[\boldsymbol{\Phi}^T\boldsymbol{\Phi}]^{-1}\boldsymbol{\Phi}^T\boldsymbol{e}\right] \\ &= \varepsilon[\boldsymbol{\theta}] + \varepsilon\left[[\boldsymbol{\Phi}^T\boldsymbol{\Phi}]^{-1}\boldsymbol{\Phi}^T\right]\varepsilon[\boldsymbol{e}] = \varepsilon[\boldsymbol{\theta}] \end{aligned}$$

so $\hat{\boldsymbol{\theta}}$ is generally biased.

Two ways to solve the problem: extended least squares and instrumental variables

To remove the correlation and the consequent bias in $\hat{\boldsymbol{\theta}}$ we must somehow modify the equation error (noise) to make it white, or modify the regressors to make them uncorrelated with the noise. The former may be done by adding a noise-structure model to (3.29) and (3.24). The *extended least-squares algorithm* uses recursive least squares with the model (3.29) extended so that

$$\begin{aligned} \boldsymbol{\Phi}_t^T &= [(-y_{t-1})\ (-y_{t-2})\ \ldots\ (-y_{t-n})\ u_{t-d}\ u_{t-d-1}\ \ldots\ u_{t-d-n+1}\ \hat{e}_{t-1}\ \hat{e}_{t-2}\ \ldots\ \hat{e}_{t-q}] \\ \boldsymbol{\Theta}^T &\equiv [a_1\ a_2 \ldots\ a_n\ b_1\ b_2 \ldots\ b_n\ c_1\ c_2 \ldots\ c_q] \end{aligned} \tag{3.51}$$

where

$$\hat{e}_{t-i} = y_{t-i} - \boldsymbol{\Phi}_{t-i}^T \hat{\boldsymbol{\Theta}}_{t-i}; \qquad i = 1, 2, \ldots, q \tag{3.52}$$

Clearly we are modelling the autocorrelated noise as a moving average of a sequence $\{e\}$. If q is large enough, we can hope that $\{e\}$ is white and hence uncorrelated with the lagged output sequences among the regressors. This algorithm is widely used and usually works well. Its asymptotic properties can be improved by a simple modification (filtering the residual sequence $\{\hat{e}\}$) but that introduces other difficulties and is not normally necessary. Other variants of least squares with the same aim exist.

The alternative, modifying the regressors, is the aim of the *instrumental variable method*. The method modifies the batch OLS estimate

$$\hat{\boldsymbol{\theta}} = [\boldsymbol{\Phi}^T \boldsymbol{\Phi}]^{-1} \boldsymbol{\Phi}^T \boldsymbol{y} \tag{3.53}$$

to

$$\hat{\boldsymbol{\theta}}_z = [\boldsymbol{Z}^T \boldsymbol{\Phi}]^{-1} \boldsymbol{Z}_T \boldsymbol{y} \tag{3.54}$$

where the error-correlated regressors making up the rows of $\boldsymbol{\Phi}^T$ have been replaced by the instrumental variables forming $\boldsymbol{Z}^T$. Since

$$\begin{aligned} plim\, \hat{\boldsymbol{\theta}}_z &= plim\, [\boldsymbol{Z}^T \boldsymbol{\Phi}]^{-1} \boldsymbol{Z}^T \boldsymbol{y} = plim\, [\boldsymbol{Z}^T \boldsymbol{\Phi}]^{-1} \boldsymbol{Z}^T (\boldsymbol{\Phi} \boldsymbol{\theta} + e) \\ &= plim\, \boldsymbol{\theta} + plim\, [\boldsymbol{Z}^T \boldsymbol{\Phi}]^{-1} \boldsymbol{Z}^T e \\ &= plim\, \boldsymbol{\theta} + plim\, [\boldsymbol{Z}^T \boldsymbol{\Phi}/N]^{-1} plim\ \boldsymbol{Z}^T e/N \end{aligned} \tag{3.55}$$

we see that $\hat{\boldsymbol{\theta}}_z$ is consistent so long as $plim\ \boldsymbol{Z}^T e/N$ is zero. (The divisions by N are necessary for the probability limits to exist.) Consistency does not imply unbiasedness in general, as $plim\, \boldsymbol{x}_N \neq lim_{N \to \infty} \mathcal{E}[\boldsymbol{x}_N]$ in general, but in many cases the limits do coincide and zero correlation between $\boldsymbol{Z}$ and e gives both consistency and unbiasedness.

The choice of $\boldsymbol{Z}$ is conditioned by the need to make it strongly correlated with $\boldsymbol{\Phi}$, which makes the covariance of $\hat{\boldsymbol{\theta}}_z$ as small as possible, while maintaining lack of correlation with e. For example, the output of an auxiliary model approximating the system being identified may be used in lieu of the noise-free system output.

3.5.7 Identification of time-varying systems

One of the most useful features of recursive estimation algorithms is their potential ability to track time-varying unknowns (parameters or, as in the Kalman filter, state variables). In the recursive least-squares-based algorithms already described, the correction gain depends on the estimated covariance P_t of the parameter estimates. As more observations are processed, the covariance falls, reducing the correction gain (provided the observations are informative about the model). This reflects the increase in confidence in the estimates as more information accumulates about their values. The algorithms' ability to adapt to new values of the parameters therefore falls, so we must modify them if we are to track time-varying parameters. Briefly, the main options for adapting recursive least-squares-type algorithms to track a time-varying system are:

- use weighted least squares with the weights $\{w\}$ making up the weighting matrix $\boldsymbol{W} = diag(w_1\ w_2 \ldots w_t)$ increasing as the time approaches the present. The commonest technique is to increase them exponentially, so

$$w_t = w_{t-1}\lambda_t, \qquad 0 \leq \lambda_t \leq 1 \tag{3.56}$$

  Here λ_t is called a *forgetting factor* and is often kept constant. However this can lead to good information being thrown away particularly if the input sequence is not sufficiently exciting;

- prevent the estimated (normalized) covariance $\boldsymbol{P}_t$ from becoming too small as t increases by resetting it to a larger value whenever its size (measured by its trace or determinant) falls below some threshold. This is called *covariance resetting*;

- using an explicit parameter-variation model. The *simple random walk* model

$$\boldsymbol{\theta}_t = \boldsymbol{\theta}_{t-1} + \boldsymbol{w}_{t-1} \tag{3.57}$$

  with $\{w\}$ white, zero-mean and of specified covariance $\boldsymbol{Q}$ often works well and is flexible enough to accommodate a wide variety of time variation. The increment-covariance $\boldsymbol{Q}$ may have to be tuned by trial and error to get a credible amount of variation in the parameter estimates if there is no strong prior evidence of how much should occur. In the *integrated random walk* model

$$\boldsymbol{\theta}_t = \boldsymbol{\theta}_{t-1} + \boldsymbol{s}_{t-1}; \qquad \boldsymbol{s}_t = \boldsymbol{s}_{t-1} + \boldsymbol{w}_{t-1} \tag{3.58}$$

  with $\{w\}$ white, $\varepsilon[\boldsymbol{w}] = 0$, $cov\ \boldsymbol{w} = \boldsymbol{Q}$. $\boldsymbol{Q}$ controls the roughness of the variation rather than its extent. This model may therefore be employed initially to obtain estimates which show the extent of the time variation, because $\boldsymbol{Q}$ has a less critical effect on the overall time variation. Once the extent is known, the simple random walk model may be used, giving a better picture of the detailed structure of the time variation. Optimal smoothing algorithms originally devised for state estimation may be applied (with some change in interpretation) to estimate earlier values of the parameters in the light of later observations; this is useful when the number of observations is small and the best possible estimates of the early values are required.

There are many other techniques for tracking, some specialized to deal, for instance, with isolated abrupt changes. Engineering judgement is needed if a suitable method is to be chosen.

3.6 MODEL-STRUCTURE SELECTION

We have looked at parameter estimation but not, so far, at how to select a suitable model structure. Although some formal techniques for model-structure selection exist and are useful, the selection process must take into account practical constraints on the model, so it is not possible to give a comprehensive account of the selection process. For example, the complexity of the model may be restricted by the need for its later inclusion in a design procedure which is inconvenient or impracticable with a more complicated model. Prior information may play a large part in determining some features of the model, and so may familiarity with a well-established but not necessarily optimal model structure. Comprehensibility is also important, and so is the availability of techniques to analyse the model (favouring linearity, for instance).

Most formal model-structure selection techniques are statistically based and rely explicitly or implicitly, on a hypothesis test. Their asymptotic properties can therefore be investigated theoretically. However, they are vulnerable to deviations of the data from the assumptions made; for instance, a small number of anomalous outliers may destroy the effectiveness of a test. The selection process usually consists in practice of selecting a class of model by largely informal reasoning, then perhaps applying a statistical test to determine the best order within that class (e.g. dynamical order in a linear, time-invariant rational transfer-function model).

Even at this last stage, it is important to look at as many aspects of the model behaviour as possible, asking whether they are satisfactory, reproducible and credible. This is model validation.

3.7 MODEL VALIDATION

For reasons similar to those outlined in Section 3.6, we shall not attempt to set out formal validation procedures. We are not able to generalize about what techniques will be useful, but the following aspects of model behaviour are often worth examination:

1. **Properties of the residuals**: whiteness (as shown by their sample autocorrelation function over the range of lags of interest), apparent stationarity when viewed by eye (particularly absence of slow drift, periodic components and sudden level changes), any systematic discrepancies such as peaks whenever the input is behaving in a particular way, and their size, usually measured by their mean-square value normalized by that of the short-term variation of the output.

2. **Estimated variances of the parameter estimates**, although these tend to be unreliable when the parameter estimates themselves are not good, which is precisely when they are most needed.

3. **Performance of the model on records, or sections of the record, other than that used to estimate the model.**

4. **Simulation-mode performance**, i.e. the output given when the model is driven by the recorded input and any earlier output (autoregressive) terms in the model are the earlier model outputs rather than the recorded output. This is often a revealing test, showing up deficiencies clearly.

5. **Values of easily interpretable quantities derived from the parameter estimates**, such as steady-state gain, time constants and the amplitudes of the modal components of a linear model's transfer function.

6. **Joint properties of the residuals and explanatory variables**, particularly cross-correlation functions: a good model should have small cross-correlations.

7. **Consistency of the results with known typical behaviour of the system**, an extension of 4, covering things like extent of time variation (e.g. annual or daily variation of environmental or biological systems), case-to-case variation, fastest/slowest features of response, amount of output variation attributed to drift (i.e. unexplained slow variation), degree of damping, dead time (pure delay).

3.8 EXPERIMENT DESIGN

For performing identification in practice we need to design experimental conditions which may play an essential role in the final result of process identification. Experiment design involves several factors. Here only some practical aspects will be discussed, with no pretence that all eventualities are covered. The most important are:

- choice of sampling interval.

- choice of input signal;

The choice of sampling interval was partly considered in Section 3.3.3, where a rate of about 10 samples per cycle (or 10 samples per settling time of a step response) has been suggested. We should also take into account that prefiltering of data is often necessary to avoid aliasing. An analog lowpass filter with the bandwidth a little smaller than the Nyquist frequency should be used before sampling the signals.

Concerning the choice of input signal we may consider several aspects. Certain identification methods require a special type of input and this is typically the case for the non-parametric models (see Sections 3.4.1 and 3.4.2). For parametric models we only require that the input is persistently exciting (p.e.), i.e. the input signal ensures adequate excitation of the dynamics. The conditions of p.e. can be interpreted in the time domain or in the frequency domain. The input sequence is p.e. of order p if no selection of every p successive samples is linearly dependent. The frequency domain counterpart is that the input contains power at p or more frequencies. An input which is p.e. of order p allows us to estimate p ordinates by least squares.

The choice of input involves also the choice of its amplitude. We should take into account the following points:

1. There are reasons for using as large input amplitude as possible to ensure sufficient SNR and hence the accuracy of the estimates;

2. On the other hand, there are also constraints on the input amplitude. For safety or economic reasons we may not be able to introduce too large fluctuations in the process. In practice most processes are non-linear. A linear model is merely an approximation which can be valid only in some region and for this reason too large an amplitude of the input should not be chosen as well.

As a summary of this section we can formulate a simple but general rule: *choose the experimental conditions that resemble the conditions under which the model will be used.*

3.9 TOPICS NOT COVERED

These include state-space, matrix fraction description and matrix difference equation models of MIMO systems; identification of systems in closed-loop conditions; specific identification methods and model structures for non-linear systems; important families of estimators such as Bayes estimators and maximum likelihood estimators; most theoretical properties of estimation algorithms; constrained estimation; model reduction; identifiability and adequacy of excitation; robust numerical techniques for ill conditioned data sets; simultaneous state and parameter estimation; criteria for tuning the model specification. Many of them are covered in an introductory way by the book by Norton [5].

3.10 LABORATORY EXERCISES

3.10.1 MATLAB software tools in laboratory course 3

Laboratory exercises are prepared to illustrate how to put into practice the identification techniques discussed in this chapter and what are the main problems we could deal with. The exercises will be done under the MATLAB software together with the System Identification (SI) Toolbox.

The SI Toolbox provides the basic help in building mathematical models of dynamical systems, based on observed system data. The SI Toolbox is a collection of m-files that implement the most common and useful parametric and non-parametric techniques available. It provides the tools for estimation and identification, covering both basic routines and more elaborate functions required for advanced applications. Details on the toolbox can be found in *System Identification Toolbox – User's Guide* by L. Ljung. Here only comments are given.

The main goal of this section is to lead the reader through the basic actions to be performed during identification. All examples presented in what follows are supported by appropriate m-files and used for tutoring in process identification and the SI Toolbox usage; however, they can also be introduced or changed by the users themselves. The examples in this section illustrate the main topics presented in the previous sections, so only a part of the function set included in the SI Toolbox is described.

The identification routines in the examples are applied for simulated data, so in the following sections the word *system* will be used for a simulated (input–output) process and the *model* stands for the result of identification. Simulation is performed by means of the SI Toolbox functions, but it can also be done by S-functions (see Section 4.8.4). The examples concern SISO model identification, though some functions for MIMO model identification are available in the SI Toolbox as well.

Using the SI Toolbox we meet principally with models, their simulation or identification. The basic model format in the SI Toolbox is the `theta` format, which is a packed matrix containing information about a model structure and its nominal or estimated parameters together with their variances. The basic way to display the information about the model is to use the `present` function. There are also some functions to transform a model from the `theta` matrix to another required form (e.g. polynomials or transfer function) and vice versa, and often they play an essential role in programs.

Other, more detailed remarks about the software tools will be given in the following sections.

3.10.2 Simulation and prediction of SISO transform models

This section introduces the reader to the SI Toolbox. It also gives an appreciation of some basic relations between parameters, poles and dynamics of a system (see Sections 3.4.3 and 3.4.4).

Example 1 – simulation of a SISO system

Declare a discrete-time system with transfer function

$$H(z^{-1}) = z^{-1}\frac{B(z^{-1})}{A(z^{-1})}$$

where

$$A(z^{-1}) = 1 - 0.9z^{-1}$$

$$B(z^{-1}) = 0.1$$

Calculate and draw a pole of the system.

Next, prepare the unit-pulse input, draw the unit-pulse response of the system (the weighting function) and its sampled-step response for $N = 50$ data.

Repeat the simulation for

$$a_1 = -0.5 \qquad b_0 = 0.5$$

and for the second-order system with

$$A(z^{-1}) = 1 - 1.5z^{-1} + 0.7z^{-2} \tag{3.59}$$

$$B(z^{-1}) = 0.1 + 0.1z^{-1}. \tag{3.60}$$

Quick help

1. Enter the program `id2_1` at the MATLAB prompt.
2. Follow the advice given in the program.
3. Shrink and move the figures so as to compare them easily.
4. Compare the poles and settling-times of the responses.
5. Try to simulate other systems according to the examples shown.

Question

What are the relations between the coefficients of polynomial $A(z^{-1})$, the poles and the character of the unit-pulse or sampled-step response?

Example 2 – prediction of the system output

For $N = 100$ data, simulate a second-order system with time delay $d = 2$, $A(z^{-1})$ and $B(z^{-1})$ as in (3.59) and (3.60). Use a random binary sequence as the system input. Generate a normal random sequence with zero mean and variance 0.01 and add the noise to the system output y_t. Compute the signal to noise ratio (SNR).

Compute and draw the one-step-ahead prediction of the system output and the prediction error based on the input sequence and known vector of system parameters.

Quick help

1. Enter the program `id2_2` at the MATLAB prompt and follow the advice given.

2. We assume that the "real" system output yv is a sum of noise-free output y and noise v. The noise v can also be passed through any dynamical filter. The SNR is defined as the ratio of the noise-free output variance to the noise-at-the-output variance.

3. In this example prediction is based on simulated system parameters. Usually, identified model parameters are used instead of unknown "real" system parameters.

4. Prediction can also be carried out for a longer horizon than for one step ahead.

Question

Derive the formulae for one-step-ahead prediction of the system output and for the prediction error.
Hint: use Eqs (3.22) and (3.25) assuming that the noise term v_t is unpredictable.

3.10.3 Model types and test signals

This section is intended to give an appreciation of some basic types of SISO models, relations between different types of models and appropriate identification methods for non-parametric models.

Non-parametric SISO time-domain models

These types of models are described in Section 3.4.1.

Example 1 – pulse testing

Generate the unit-pulse response of the second-order system given by (3.59) and (3.60) with time delay $d = 2$.

Now, we are going to take into account the presence of noise in the system output. Generate a normal random sequence with zero mean and variance 0.0001 and draw the unit-pulse response with added noise.
Repeat it for noise variance 0.01.

Quick help

1. Enter the program `id3_1` at the MATLAB prompt and follow the advice given.

2. The "real" unit-pulse response of the system y is the sum of noise-free response and noise. You can observe how much noise at the output influences the unit-pulse response.

Question

What are the drawbacks of unit-pulse testing?

Example 2 – sampled-step testing

Simulate the system as in Example 1, but now put $u_i = 1, i \geq 1$. Add the noise to the output as in the previous case.

Quick help

1. Enter the program `id3_2` at the MATLAB prompt and follow the advice given.

2. The "real" sampled-step response of the system y is the sum of noise-free response and noise. You can observe how much noise at the output influences the sampled-step response.

Question

What are the drawbacks of sampled-step testing?

Example 3 – correlation-based identification of the unit-pulse response

The need to minimize the effects of noise suggests that other input signals and processing of input–output measurements should be employed.

Generate a random binary sequence as an input u_t and simulate the system output y_t with added noise for $N = 1000$ measurements. Calculate the autocorrelation function $r_{uu,k}$ for the input sequence u_t and maximum lag $k_{max} = 50$.
Calculate the cross-correlation function $r_{uy,k}$ for 1000 data and repeat it for only 100 data.

Quick help

1. Enter the program `id3_3` at the MATLAB prompt and follow the advice given.

2. We assume that the "real" system output y is the sum of noise-free response to the input sequence and noise. To get a better SNR, the input sequence has greater power than the unit pulse while the noise power is the same.

3. To draw more general conclusions you are recommended to repeat the simulation for other input sequences, noise variances and numbers of data.

Questions

1. Compare the cross-correlation function $r_{uy,k}$ with the true unit-pulse response.

2. What are the advantages of correlation analysis in unit-pulse response identification?

3. Why are random binary sequences convenient in correlation analysis?

4. How do the results depend on the number of data?

Non-parametric SISO frequency-domain models

The technique of Fourier transfer-function estimation is presented. Sine-wave testing is omitted because of its disadvantages (see Section 3.4.2). Example 4 is concerned with the Fourier transform, while Example 5 introduces estimation of discrete Fourier transfer functions.

Example 4 – discrete Fourier transform

Let $x(t)$ be a deterministic signal. If its energy is finite

$$\int_{-\infty}^{\infty} x^2(t)dt < \infty \tag{3.61}$$

then a continuous Fourier transform $X(j\omega)$ of $x(t)$ exists and is

$$X(j\omega) = \int_{-\infty}^{\infty} x(t)e^{-j\omega t}dt \tag{3.62}$$

where ω is angular frequency [rad/s].

Frequency f, expressed in Hz, is connected to angular frequency by $\omega = 2\pi f$.

For discrete-time series x_t, obtained by sampling $x(t)$ at equal time T, the discrete Fourier transform can be written in the form

$$X(j\omega T) = T \sum_{n=-\infty}^{\infty} x_n e^{-j\omega Tn} \tag{3.63}$$

If a data sequence x_n is available only for finite time instants $n = 0, 1, \ldots, N-1$ and the frequency scale is discretized also into N values, then the discrete Fourier transform (DFT) is defined as

$$X^N(j\omega T) = T \sum_{n=0}^{N-1} x_n e^{-j\omega T n} \tag{3.64}$$

where $\omega T = \frac{2\pi k}{N}$ for $k = 0, 1, \ldots, N-1$.

A computer realization of the (3.64) is known as the fast Fourier transform (FFT).

Calculate the DFT $X^8(j\omega T)$ of the data sequence $x_t = 0, 1, 0, 0, 0, 0, 0, 0$ and draw:

1. the real part and the imaginary part of $X^8(j\omega T)$ as functions of ωT;
2. the amplitude and the phase of $X^8(j\omega T)$ as functions of ωT;
3. $X^8(j\omega T)$ in polar coordinates.

Quick help

Enter the program `id3_4` at the MATLAB prompt; it computes and plots the Fourier transform obtained by using the `fft(x)` function.

Question

Compare your results with the MATLAB outcomes.

Example 5 – empirical transfer-function estimator

A stable, linear discrete-time SISO system is considered

$$y_t = \sum_{i=1}^{\infty} h_i u_{t-i} + v_t \tag{3.65}$$

The empirical transfer-function estimator (ETFE) approach aims at determining estimates of $H(j\omega T)$ at some or at all equally spaced frequencies $\omega T = \Omega n$. $\Omega = \frac{2\pi}{N}$ is the fundamental frequency and $n = 0, 1, \ldots, N/2$ denote consecutive harmonics of this frequency up to the Nyquist frequency. $H(j\Omega n)$ should be determined on the basis of an N-sample input data sequence u_t and output data sequence y_t ($t = 0, 1, \ldots, N-1$). The estimator is defined as the ratio of Fourier transforms

$$\hat{H}(j\Omega n) = \frac{Y^N(j\Omega n)}{U^N(j\Omega n)} \tag{3.66}$$

where $U^N(j\Omega n)$ and $Y^N(j\Omega n)$ are discrete Fourier transforms of the input and output data sequences respectively.

ETFE can be calculated by segmented or non-segmented data processing:

- the essence of segmented identification is to identify k models for consecutive N-sample output data sequences separately and average the models.
  The sth model ($s = 0, 1, \ldots, k-1$) is estimated by

$$\hat{H}(j\Omega n, s) = \frac{Y^N(j\Omega n, s)}{U^N(j\Omega n, s)} \tag{3.67}$$

  where

$$U^N(j\Omega n, s) = \sum_{n=0}^{N-1} u_{n+sN} e^{-j\Omega n} \tag{3.68}$$

$$Y^N(j\Omega n, s) = \sum_{n=0}^{N-1} y_{n+sN} e^{-j\Omega n} \tag{3.69}$$

  and the overall model is specified as

$$\overline{\hat{H}}(j\Omega n) = \frac{1}{k} \sum_{s=0}^{k-1} \hat{H}(j\Omega n, s) \tag{3.70}$$

- the essence of non-segmented identification is to identify a single model for the entire kN-sample output data sequence by

$$\hat{H}(j\Omega n) = \frac{Y^{kN}(j\Omega n)}{U^{kN}(j\Omega n)} \tag{3.71}$$

Let us consider a discrete-time system described by its transfer function

$$H(z^{-1}) = z^{-1} \frac{0.6}{1 - 0.8z^{-1}} \tag{3.72}$$

Excite the system by a sequence of $N = 512$ samples of a random independent $\mathcal{N}(0, 1)$ variable. Assume that $v_t \equiv 0$. Identify the frequency characteristics based on non-segmented processing.

Repeat ETFE identification employing the segmented method $k = 4$ segments and $N = 128$.

Quick help

1. Enter the program `id3_5` at the MATLAB prompt and follow the advice given.
2. ETFE is evaluated by means of the `etfe` function, which also provides some smoothing of the crude estimate.
3. The data come from simulation, so we assume by default that the sampling interval $T = 1$.

4. ETFE is plotted together with the true system transfer function, obtained from `theta` format by means of the `th2ff` function.

5. Frequency transfer functions are displayed as amplitude and phase plots with linear frequency scales and Hz as the frequency unit. They may be also shown in Bode diagram form by means of the `bodeplot` function, or in Nyquist diagram form using the `nyqplot` function.

Question

1. Compare the results obtained by both methods and give some conclusions.

2. Try to write a program evaluating transfer function (3.66) based directly on the `fft(x)` function, without using `etfe`.

Parametric SISO time-domain and transform models

Now we are concerned with the SI Toolbox implementation of models discussed in Section 3.4.4.

Example 6 – equation-error model

An equation-error model given by a linear regression equation

$$y_t = -\sum_{i=1}^{nA} a_i y_{t-i} + \sum_{i=1}^{nB} b_i u_{t-d-i+1} + v_t = \boldsymbol{\Phi}_t^T \boldsymbol{\Theta} + v_t \tag{3.73}$$

or in the transform notation

$$y_t = z^{-d} \frac{B(z^{-1})}{A(z^{-1})} u_t + \frac{1}{A(z^{-1})} v_t \tag{3.74}$$

is often called an *ARX-model* (AutoRegressive with eXogenous input).

For a given model estimate $\hat{\boldsymbol{\Theta}}$, the one-step-ahead prediction of the output is

$$\hat{y}_t = \boldsymbol{\Phi}_t^T \hat{\boldsymbol{\Theta}} = (1 - \hat{A}(z^{-1}))y_t + z^{-d}\hat{B}(z^{-1})u_t$$

and the simulated model output is given by

$$y_t^M = z^{-d} \frac{\hat{B}(z^{-1})}{\hat{A}(z^{-1})} u_t$$

Simulate and identify the equation-error model. Plot the prediction of the output and simulated model output based on the estimated model.

Quick help

1. Enter the program `id3_6` at the MATLAB prompt and follow the advice given.

2. To obtain the input–output data sequences, the ARX-system (3.73) is simulated, assuming noise v_t influences the system. The model of the system is estimated using the least-squares algorithm presented in Section 3.10.4.

3. The ARX-model structure in the *SI Toolbox* is defined as [nA nB d].

4. Information about the model is stored in `theta` format. Full format description can be obtained using the command `help theta`.

5. The function `present` is used to display information about the identified model packed in `theta`. It presents estimated polynomial parameters of the model, together with their estimated standard deviations and how the model was created.

Question

Show the differences between the prediction of the output and the simulated model output.

Example 7 – output-error model

In contrast to the equation-error model (3.73), in the output-error (OE) model it is assumed that the noise v_o does not pass through any of the system's dynamics, so

$$y_t = z^{-d}\frac{B(z^{-1})}{A(z^{-1})}u_t + v_{ot} \tag{3.75}$$

This formula is non-linear in the coefficients in $A(z^{-1})$.

Simulate and identify the output error model. Plot the prediction of the output and the simulated model output based on the estimated model.

Quick help

1. Enter the program `id3_7` at the MATLAB prompt and follow the advice given.

2. To obtain the input–output data sequences, the OE-system (3.75) is simulated. The OE-model of the system is estimated using a prediction error algorithm, which is a modified version of extended least-squares, presented in sec. 3.10.4.

3. The OE-model structure in the *SI Toolbox* is defined as [nB nA d].

4. Information about the model is stored in `theta` format. Function `present` is used to display information about the identified model.

Question

Show that the prediction of the output is equivalent to the simulated model output for the OE-model.

3.10.4 Identification algorithms

This section is intended to show some properties of the most popular algorithms in parametric model identification (see Section 3.5). We shall also examine some basic model validation techniques.

Batch ordinary least-squares estimation

Batch ordinary least-squares (OLS) (presented in Section 3.5.1) is one of the most popular and useful identification methods applied for parameter estimation of equation-error models (ARX-models). However, for certain conditions, when the regressors are strongly interrelated, OLS can give unacceptable results. Example 1 shows the influence of the input signal. The effects of model order are presented in Example 2, while the influence of SNR and number of data is examined in Example 3.

Example 1 – choice of the input signal

The results of OLS estimation depend on the proper choice of the input signal (see discussion on persistent excitation in Section 3.8). The estimation can fail if the OLS normal equation is ill conditioned.

For $N = 1000$ data points and sampled-step input, simulate the discrete-time SISO system with the transfer function:

$$H(z^{-1}) = z^{-2}\frac{B(z^{-1})}{A(z^{-1})}$$

where $A(z^{-1})$ and $B(z^{-1})$ are given by (3.59) and (3.60).

Next, using the OLS method, estimate the parameters of an ARX-model with the same structure as the simulated system.

Afterwards, replace the sampled step in the input by a random binary sequence and for new data try to identify a model again.

Quick help

1. Enter the program `id4_1` at the MATLAB prompt and follow the advice given.
2. To draw more general conclusions, it is recommended that you repeat simulation for other input sequences (e.g. sum of sinusoids, random signal, unit pulse).

Question

Why was the identification wrong in the first case and why did matrix-rank deficiency occur? Give some rules for the choice of input.

Example 2 – choice of model order

In parametric model identification procedure the model structure (i.e. polynomial orders and time delay) must be declared. The proper choice of model order has an essential influence on OLS results (see Sections 3.6 and 3.7).

Identify ARX models with $nA = nB = 1$ and $nA = nB = 3$, based on the data sequences obtained in Example 1 with random binary input.

Simulate noise at the output as $v_{ot} = \frac{1}{A(z^{-1})}v_t$, where v_t is a random independent sequence with variance $\sigma_v^2 = 0.01$, and add the noise to the output. Estimate the model parameters for three orders, as before.

Compute the one-step-ahead prediction $\hat{y}_t$ and prediction error e_t based on the identified models.

Compute the simulated model output y_t^M.

Quick help

1. Enter the program `id4_2` at the MATLAB prompt and follow the advice given.
2. Simulation of the system with output corrupted by noise is done by individual generation of noise-free output and noise at the output. Such a simulation of ARX-system may be done in one step by using the `idsim` function with a specified matrix of sequences of input and noise `[u v]`.

Questions

1. In what cases it is impossible to solve the OLS normal equation? Why?
2. Compare the simulated system parameters with the estimated model parameters together with their standard deviations.
3. Write proper formulas for one-step-ahead prediction, prediction error and simulated model output for all identified model structures.
4. Compare the system output y_t with its one-step-ahead prediction $\hat{y}_t$ in all three cases.
5. Compare the variances of the prediction error with variance σ_v^2. How good can one-step-ahead prediction be?
6. Compare the system output y_t with model output y_t^M for three models.
7. Compare the variances of difference between system output and simulated model output with the variance of the disturbance added to the output v_o. How good can simulation of the model output be?
8. How does the variance of the prediction error change as the model order increases?

Example 3 – influence of number of data and SNR on parameter estimation

For a simulated system with added noise as in Example 2, compute the signal to noise ratio (SNR).

Repeat the identification for correct model structure, but for only $N = 100$ data.

Next, simulate the same system, but insert noise having 100 times greater variance. Compute the SNR and identify a second-order model for $N = 1000$ and 100 data.

Draw the one-step-ahead prediction $\hat{y}_t$ and the simulated model output y_t^M.

Quick help

1. Enter the program `id4_3` at the MATLAB prompt and follow the advice given.
2. Simulation of the system with output corrupted by noise is done by individual generation of noise-free output and noise at the output according to the SNR calculation. Simulation of an ARX-system may also be done in one step by using the `idsim` function with specified matrix of sequences of input and noise `[u v]`.
3. The variances of the prediction error and the difference between system output and simulated model output are presented to see the relationships between them and noise v and noise at the output v_o.
4. To draw more general conclusions, it is worth repeating simulation and identification several times because of the randomness of the results.

Questions

1. Compare the simulated system parameters with the estimated model parameters.
2. Compare the standard deviations of the estimated parameters for short and long data sequences, for the large and small SNR.
3. Compare the variances of:
   - the prediction error e_t with the simulated noise v_t;
   - the difference between simulated output y_t^M and output y_t with noise at the output v_{ot}.

   Draw conclusions.
4. How do the number of measurements N and the SNR influence the results?

Recursive ordinary-least-squares estimation

The RLS algorithm is an alternative to LS identification, doing estimation recursively, in contrast to batch OLS where estimates are computed from all the data in one go (see Section 3.5.3). Here only basic features of RLS are presented. The ability to track time-varying systems is discussed in Section 3.5.7.

Example 4 – comparison with OLS

For $N = 1000$ data points and random binary input, simulate the ARX-system with polynomial $A(z^{-1})$ given by (3.59), polynomial $B(z^{-1}) = 0.2 + 0.1z^{-1}$, delay $d = 2$ and noise at the output with variance $\sigma_v^2 = 0.01$.

From the generated data, identify the ARX-model using the OLS and RLS algorithms.

Display the time sequences of estimated RLS parameters.

Quick help

1. Enter the program `id4_4` at the MATLAB prompt and follow the advice given.
2. Function `rarx` is used for off-line RLS application, where first all the data are collected and then a model is identified. This function may be also employed in true on-line applications, so that estimation is carried on at the same time as the data become available.
3. The estimated parameters are stored in a matrix, of which row k contains the parameters associated with time k. In program `id4_4` only the last estimates ($k = N$) are displayed, although the others can also be presented.
4. The plots of parameter estimates show both true (i.e. simulated) and estimated model parameters to see how they converge.
5. The RLS algorithm is used without weighting the measurements, i.e. with forgetting factor $\lambda = 1.0$. Default initial values are zeros for parameters θ_0, and diagonal covariance matrix $\boldsymbol{P}_0$ with values of 10^4 on diagonal.

Questions

1. Compare the results obtained for both LS algorithms.
2. How do the number of data N and initial value of $\boldsymbol{P}_0$ influence the RLS results?

Algorithm to avoid bias due to correlation between regressors and noise

The OLS estimate is unbiased if the noise is zero-mean and independent of the regressors. If these conditions are not fulfilled, the equation error (noise) ought to be modified to make it white. Another way to remove the correlation and the consequent bias in $\hat{\boldsymbol{\Theta}}$ is based on modification of the regressors to make them uncorrelated with the noise. Both methods are presented in Section 3.5.6.

Example 5 – extended least-squares

The *extended least-squares* (ELS) algorithm uses recursive least squares with an extended model

$$\hat{\boldsymbol{\Theta}}^T = [a_1\ a_2 \ldots\ a_{nA}\ b_1\ b_2 \ldots\ b_{nB}\ c_1\ c_2 \ldots\ c_{nC}]$$

$$\boldsymbol{\Phi}_t^T = [(-y_{t-1})\,(-y_{t-2})\,\ldots\,(-y_{t-nA})\;u_{t-d}\;u_{t-d-1}\,\ldots\,u_{t-d-nB+1}$$
$$\hat{e}_{t-1}\;\hat{e}_{t-2}\ldots\,\hat{e}_{t-nC}]$$

where

$$\hat{e}_{t-i} = y_{t-i} - \boldsymbol{\Phi}_{t-i}^T \hat{\boldsymbol{\Theta}}_{t-i}; \qquad i = 1, 2, \ldots, nC$$

Such a model type is called an ARMAX-model (AutoRegressive Moving Average with eXogenous input).

A modified variant of the ELS algorithm, called RPE (Recursive Prediction Error), is used in the SI Toolbox.

Declare the ARMAX-system given by the polynomials (3.59), (3.60) and by

$$C(z^{-1}) = 1 + 0.8z^{-1}$$

with delay $d = 2$.

Generate the system output sequence for $N = 1000$ data, random binary input and random noise at the output with variance $\sigma_v^2 = 0.01$.

From the received data, identify an ARX-model with structure [2 2 2] and ARMAX-model with structure [2 2 1 2].

Test the residuals of both models.

Quick help

1. Enter the program `id4_5` at the MATLAB prompt and follow the advice given.

2. In applied RPE estimation, a quadratic prediction error criterion is minimized using an iterative Gauss–Newton algorithm. The outcomes from all iterations are displayed.

3. The ARMAX-model structure in the *SI Toolbox* is defined as [nA nB nC d].

4. The residuals are tested employing the `resid` function, which shows the autocorrelation function of the residuals (prediction errors) and the cross-correlation function between residuals and input together with 99% confidence intervals. The test is based on the assumption that the residuals ought to be white (uncorrelated) and independent of the input. If the computed values of correlation lie inside the confidence interval, we can assume that this test is satisfied.

Questions

1. Compare the results obtained for ARX and ARMAX model identification.

2. Why does the identification of the ARX-model using the OLS method fall down in this case?

Example 6 – instrumental variables

The alternative, modifying the batch OLS estimate, is the aim of the instrumental variable (IV) method. Here

$$\hat{\Theta}_z = [Z^T \Phi]^{-1} Z^T y$$

The choice of Z is conditioned by the need to make it strongly correlated with Φ, while maintaining lack of correlation with error e. For example, the output of a model approximating the system being identified may be used in lieu of the noise-free system output.

For the system simulated in Example 5, identify an ARX-model using the OLS method. Then, applying the known input sequence and estimated model, generate the model output sequence and put it into the IV algorithm. Repeat the procedure several times using the new model.

Simulate an OE system with polynomials $A(z^{-1})$ and $B(z^{-1})$ as before and add the noise at the output with variance $\sigma^2_{v_o} = 1$. Identify an ARMAX-model with structure [2 2 2 2] and an ARX-model using the IV method.

Quick help

1. Enter the program `id4_6` at the MATLAB prompt and follow the advice given.

2. Simulated model outputs are generated as the instrumental variables. The first instruments are based on the model obtained via the OLS algorithm.

3. The IV algorithm does not return any estimated covariance matrix for the parameters `theta`.

Questions

1. Compare the results of ARX-model identification employing the OLS and IV methods.

2. Compare the estimates of polynomials A and B for the ARMAX- and ARX-models using the IV algorithm.

3. How does the number of data influence the results of IV estimation? To answer this question, use the same program modifying the number of data N.

4. Try to propose another way of generating instrumental variables.

Identification of time-varying systems

Example 7 – RLS estimation with a forgetting factor

One of the most useful features of recursive estimation algorithms is their potential ability to track time-varying system. The main option for adapting RLS algorithms to track time-varying parameters is to use weighted least squares with a *forgetting factor* (see

3.5.7).

Use the MATLAB program `id4_7` to simulate the second-order system with varying parameters of polynomial $A(z^{-1})$

$$a_1 = -1.5; \quad a_2 = 0.7 \qquad for\ t \leq 100;$$
$$a_1 = -1.2; \quad a_2 = 0.9 \qquad for\ t > 100;$$

while

$$B(z^{-1}) = 0.2 + 0.1z^{-1}$$

and delay $d = 2$.

Subsequently, identify an ARX-model with the correct structure using the RLS algorithm with forgetting factor $\lambda = 1.0$ and draw time-series of model parameter estimates.

Repeat the identification for $\lambda = 0.95$ and $\lambda = 0.9$.

Quick help

1. Enter the program `id4_7` at the MATLAB prompt and follow the advice given.
2. If the diagrams are not clear enough, enlarge them to see the differences between the estimates obtained using different forgetting factors.

Questions

1. How does the forgetting factor affect noise damping and tracking of time-varying parameters? Give an idea of how to select the forgetting factor.
2. Try to explain the influence of the forgetting factor on the tracking capabilities .

3.10.5 Model structure selection and model validation

The identification procedure typically ends up with a collection of models with different structures, so we have to decide which model is best and if it is an adequate model for specified purposes. These are the problems of *model validation* and *model structure selection*, which constitute the common last step of process identification (see Sections 3.6 and 3.7).

Although some formal techniques for model selection exist, the selection must also take into account practical constraints on the model, so it is not possible to generalize about what techniques will be useful. The following aspects of model behaviour are often worth examination:

- properties of the residuals;
- performance of the model on records, or sections of the record, other than that used to estimate the model (cross-validation);
- positions of model poles and zeros.

Example 1 – testing the residuals

The residuals (prediction errors) e_t should be white and independent of the input u_t. Thus, analysing the autocorrelation function of the residuals or their cross-correlation function with the input, we are able to test the model.

Simulate the second-order system (see Example 1 in Section 3.10.4) with random binary input.

Identify ARX-models with three different structures (e.g. [1 1 2],[2 2 2],[3 3 2]), compute and test the residuals.

Quick help

1. Enter the program `id5_1` at the MATLAB prompt and follow the advice given.
2. The residuals are tested employing the `resid` function, which shows the autocorrelation function of residuals and the cross-correlation function between residuals and input, together with 99% confidence intervals. If the computed values of correlation lie inside the confidence interval, we can assume that this test is satisfied.
3. If the plots are not clear enough, enlarge them to check if the bounds of the confidence intervals are exceeded.

Questions

1. Compare the displayed autocorrelation and cross-correlation functions.
2. Try to formulate a rule for testing whiteness in model validation.
3. Is it possible to use a test of whiteness to detect the overparameterized models (i.e. having higher order than necessary)?

Example 2 – cross-validation

Employment of two data sets for identification:

- the *estimation data set* used only for model parameter estimation;
- the *validation data set* used only for model validation

can be a powerful tool to check the correctness of the model.

Divide the data set generated as in Example 1 into two subsets. Then, using the first data subset, identify three models with different structures. Employing the second data subset, compute the one-step-ahead prediction of the output based on the identified models, display their time-series together with the system output and compute the prediction error variances.

Quick help

1. Enter the program `id5_2` at the MATLAB prompt and follow the advice given.
2. the data set is divided into half sets. This division may be done in any other way; the only restriction is that the system parameters must not vary in time.
3. Only the one-step-ahead predictions for different model structures are tested on the validation data set. The whiteness of residuals or the simulated model output may be also tested in this way. Try it.
4. If diagrams are not clear enough, enlarge them to see how good prediction of the output is for different models.

Questions

1. Compare the time series and prediction error variances.
2. Give an idea how to test the model by cross-validation.

Example 3 – checking pole-zero cancellation

Sometimes, testing the values of model poles and zeros, we are able to decide that the model order is too high, assuming that some pole-zero pairs may be cancelled.

For models evaluated as in Example 1, calculate and plot their poles and zeros.

Quick help

1. Enter the program `id5_3` at the MATLAB prompt and follow the advice given.
2. To decide if any pole is sufficiently near a zero for this pair to be cancelled seems rather subjective and not very formal, so this test should be used in connection with other tests. Enlarge the plots if they are not clear enough.

Question

How should we examine the model order by testing poles and zeros?

3.11 REFERENCES

[1] K. J. Åström and P. Hagander and J. Sternby. Zeros of sampled systems. *Automatica*, **20**, 31–38, 1984.

[2] G. J. Bierman. *Factorization Methods for Discrete Sequential Estimation.* Academic Press, New York, 1977.

[3] L. Ljung. *System Identification. Theory for the User.* Prentice Hall, Englewood Cliffs, NJ, 1987.

[4] R. M. Middleton and G. C. Goodwin. *Digital Control and Estimation: a Unified Approach.* Prentice Hall, Englewood Cliffs, NJ, 1990.

[5] J. P. Norton. *An Introduction to Identification.* Academic Press, London, 1986.

[6] T. Söderström and P. Stoica. *System Identification.* Prentice Hall, Englewood Cliffs, NJ, 1989.

[7] P. E. Wellstead and M. B. Zarrop. *Self-Tuning Systems. Control and Signal Processing.* Wiley, Chichester, 1991.

4

Neural networks in identification and control

4.1 INTRODUCTION

The topic of artificial neural networks (ANNs) for identification and control is at present one of the key research areas in the field of control systems. ANNs have been proposed by information and neural science as a result of the study of the mechanisms and structures of the brain. This has led to the development of new computational models, based on this biological background, for solving complex problems like pattern recognition, fast information processing, learning and adaptation.

4.1.1 Origins of connectionist research

In the early 1940s the pioneers of this field [22] studied the potential and capabilities of the interconnection of several basic components based on the model of a neuron. Others, like Hebb [9], were concerned with the adaptation laws involved in neural systems. Rosenblatt [28] coined the name Perceptron and devised an architecture which has subsequently received much attention. Minsky and Papert [24] introduced a rigorous analysis of the Perceptron. In the 1970s the work of Grossberg [8] came to prominence. This work, based on psychological evidence, proposed several architectures of non-linear dynamical systems with novel characteristics. Hopfield [11] applied a particular dynamical structure to solve technical problems like optimization. In 1986 the parallel distributed processing group published a series of results and algorithms [29]. This work gave a strong impulse to the area and provided the catalyst for much of the subsequent research in this field. An excellent collection of the key papers in the development of the models of neural networks can be found in Anderson and Rosenfeld [4]. Many examples of real-world applications ranging from finance to aerospace are currently being explored [10].

4.1.2 Neural networks and control

In order to provide a rational assessment of new methods it is essential to compare the emerging technologies with well-established and traditional techniques. With reference to ANNs in control and identification, the following characteristics and features are important:

- Neural network have greatest promise in the realm of non-linear control problems. This is implied by their theoretical ability to approximate arbitrary non-linear mappings. Networks may also achieve more parsimonious modelling than alternative approximation schemes, e.g. based on polynomials.

- Neural networks have a highly parallel structure which lends itself immediately to parallel implementation. Such an implementation can be expected to achieve a higher degree of fault tolerance and speed of operation than conventional schemes. Furthermore, the elementary processing unit in a neural network has a very simple structure. This also results in an increase of the processing speed.

- Hardware implementation is closely related to the preceding point. Not only can networks be implemented in parallel, a number of vendors have recently introduced dedicated VLSI hardware implementation. This brings additional speed and increases the scale of networks which can be implemented.

- As far as learning and adaptation is concerned, the networks are trained using past data records from the system under study. A suitably trained network has the ability to generalize when presented with inputs not appearing in the training data. Network can also be adapted on-line.

- Data fusion. Neural networks can operate simultaneously on both quantitative (numerical) and qualitative (symbolic) data. In this aspect networks stand somewhere in the middle ground between traditional engineering systems (quantitative data) and processing techniques from artificial intelligence (symbolic data).

- Neural networks naturally process many impulses and have many outputs; they are readily applicable to multivariable systems.

From the control systems viewpoint the ability of neural networks to deal with non-linear systems is perhaps most significant. The networks are used to provide the non-linear system models required by the techniques for synthesis of non-linear controllers. The neural networks based methods have an immense value for design of non-linear adaptive controllers for dynamical systems with poorly known and "difficult" dynamics. The learning algorithms are directly applicable as controller strategies for these controllers. Exactly the same arguments are applicable to non-linear identification which is considered here as a component of a general area of control systems but not as a separate field (in contrast to a traditional setting). Excellent text on such application can be found in recent special issues of *IEEE Control Systems* (1988, 1989, 1990, etc.) entirely dedicated to the topic discussed here. The compilation books [23, 34] provide a broad overview of the field of neural networks in control and identification. Some recent textbooks deliver fine and well-structured introductions to the general field of ANNs.

4.2 NETWORK ARCHITECTURE

The network architecture is defined by the basic processing elements and the way in which they are interconnected.

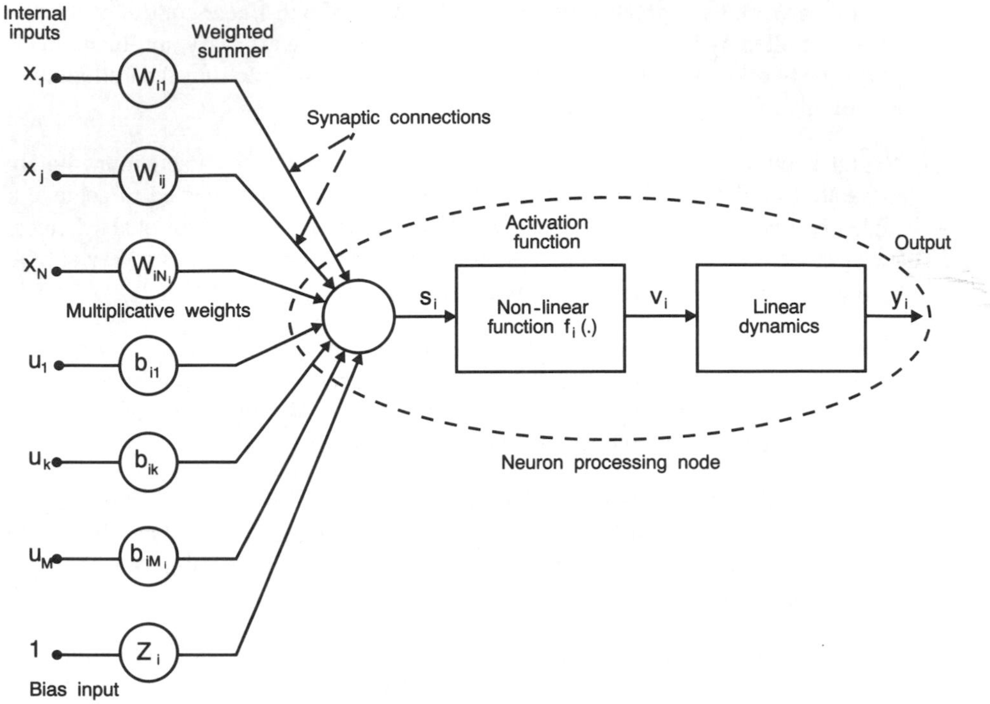

Figure 4.1. Basic model of neuron

4.2.1 Neurons

The basic processing element of the connectionist architecture is often called a neuron by analogy with neurophysiology, but other names such as perceptron [28] or adaline [35] are also used. The basic model of a neuron is illustrated in Fig. 4.1. The neuron is composed of three components:

- A weighted summer $\sum$.
- A linear dynamical single-input, single-output (SISO) system.
- A non-dynamical, non-linear function which is also called the activation function.

The weighted summer is described by

$$s_i(t) = \sum_{j=1}^{N_i} w_{ij} x_j(t) + \sum_{k=1}^{M_i} b_{ik} u_k(t) + z_i \tag{4.1}$$

giving a weighted sum s_i in terms of the internal inputs x_i, external (control) inputs u_k and corresponding weights w_{ij} and b_{ik} together with constants z_i which play a role of standard bias; t denotes a time variable which can be either continuous or discrete.

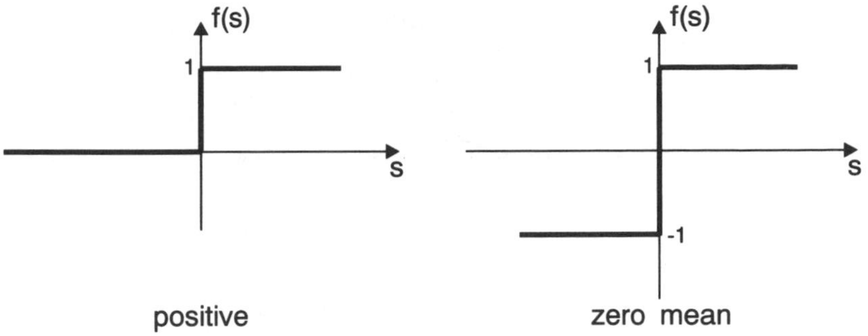

Figure 4.2. Threshold activation function

Equation (4.1) can be written in a matrix form as

$$s_i(t) = \boldsymbol{W}_i \boldsymbol{x}(t) + \boldsymbol{B}_i \boldsymbol{u}(t) + z_i \tag{4.2}$$

The linear dynamical system has input v_i and output y_i. The variable y_i is the ith neuron output. Its mathematical model can be written for continuous systems as

$$T_i y_i(t) + y_i(t) = v_i(t) \quad \text{(an inertia term)} \tag{4.3}$$

or more generally as

$$\alpha_0 y_i(t) + \alpha_1 y_i(t) = v_i(t). \tag{4.4}$$

The discrete-time model can be represented as

$$T_i y_i(t+1) + (1 - T_i) y_i(t) = v_i(t). \tag{4.5}$$

The non-dynamical non-linear function $f_i(\cdot)$ (activation function) gives the signal $v_i(t)$ in terms of the summer output $s_i(t)$:

$$v_i = f_i(s_i). \tag{4.6}$$

There are different activation functions and their selection depends on the case under consideration. For example, in pattern recognition a threshold function (see Fig. 4.2) is typically chosen, while in identification or control design the sigmoid and radial functions (see Figs 4.3 and 4.4) seem to be more popular.

4.2.2 Statical multi-layer feed-forward networks

If $T_i = 0$ in Eqs (4.3) and (4.5) for all neurons, then the network is statical. The N statical neurons in parallel fed by the input vector $\boldsymbol{x} = (x_1, \ldots, x_L)^T$ and producing the output vector $\boldsymbol{y} = (y_1, \ldots, y_N)^T$ constitute a one-layer feed-forward network which is illustrated in Fig. 4.5. The one-layer network implements a statical non-linear mapping relating the input $\boldsymbol{x}$ and the output $\boldsymbol{y}$ as (see Fig. 4.6)

$$\boldsymbol{y} = \boldsymbol{F}(\boldsymbol{W}\boldsymbol{x}) \tag{4.7}$$

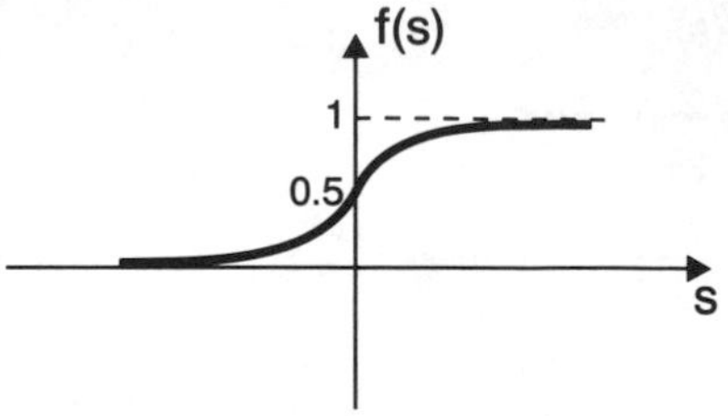

Figure 4.3. Sigmoid activation function

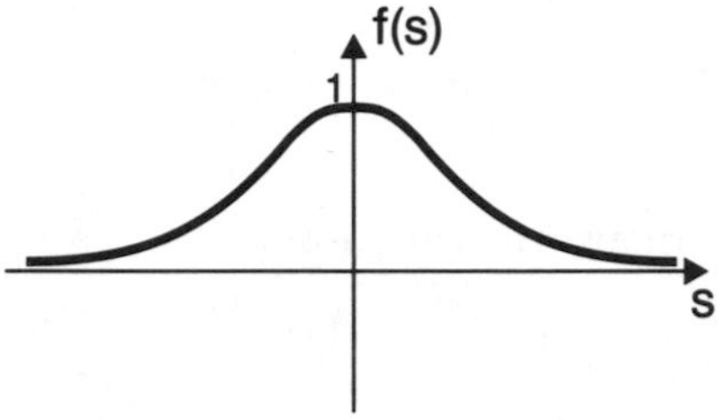

Figure 4.4. Radial activation function

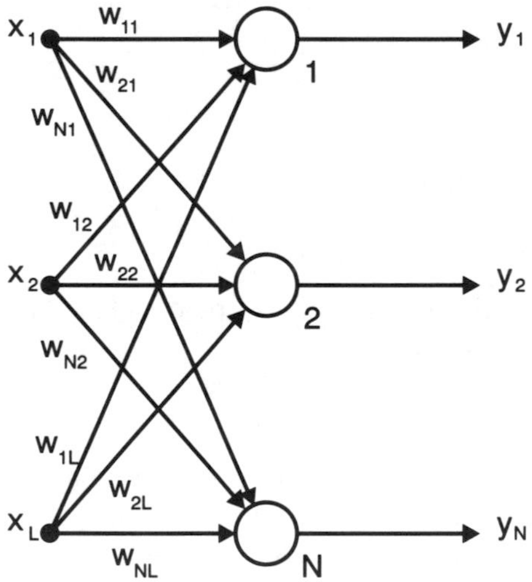

Figure 4.5. One-layer feed-forward network

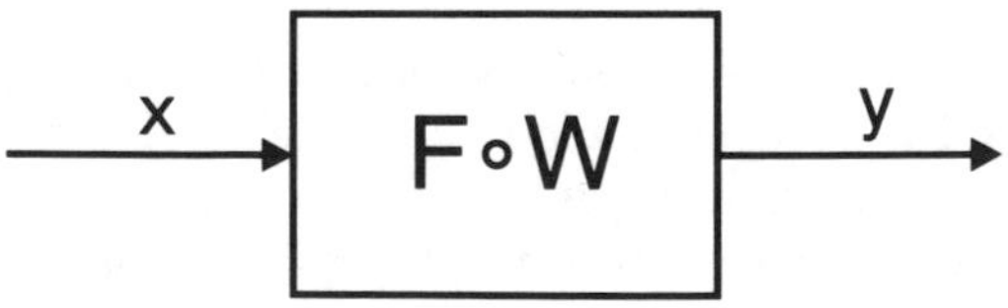

Figure 4.6. Non-linear mapping of a network

where

$$W = \begin{bmatrix} w_{11} & w_{12} & \cdots & w_{1L} \\ \vdots & \vdots & \ddots & \vdots \\ w_{N1} & w_{N2} & \cdots & w_{NL} \end{bmatrix} \tag{4.8}$$

and

$$F = \text{diag}(f_1(s_1), \ldots, f_N(s_N)). \tag{4.9}$$

Therefore, the network together with control inputs can be described with two mappings (linear and non-linear) as

$$y = F(s)$$

and

$$s = Wx + Bu + z \tag{4.10}$$

or

$$y = F(Wx+Bu + z).$$

Connecting the three layers in a cascade we obtain a two-layer feed-forward neural network which is illustrated in Fig. 4.7. The first layer is called the input layer and is composed of three neurons. The input layer output vector $y^1 = (y_1, y_2, y_3)^T$ is an input to the next layer, called the hidden layer, which is composed of two neurons. The third layer, called the output layer, produces the network output vector $y^3 = (y_6, y_7, y_8)^T$. The output layer is fed by the hidden layer output vector $y^2 = (y_4, y_5)^T$.

The network implements a non-linear mapping relating input x and output y^3 in the following composed way:

$$y^3 = [F^3 [W^3 F^2 [W^2 F^1 [W^1 x]]]]$$

where F^1, F^2, F^3 and W^1, W^2, W^3 are the matrices of activation functions and weights of the layers, respectively.

Approximation properties of the feed-forward networks

It has recently been shown by, e.g., Hornik *et al.* [12] using the Stone–Weierstrass theorem, that a two-layer network with a suitable number of nodes in the hidden layer can approximate any continuous function $h(\cdot) \in C(R^L, R^N)$, over a compact subset of R^L.

This implies that feed-forward neural networks with even one hidden layer are adequate for purposes of characterization. Since polynomials and orthogonal expansions can also approximate functions in $C(R^L, R^N)$ to any degree of accuracy, the advantages of neural networks over such representations are less obvious and have to be justified on the basis of practical considerations. In particular, the following questions have to be addressed if representations using neural networks are to be preferred:

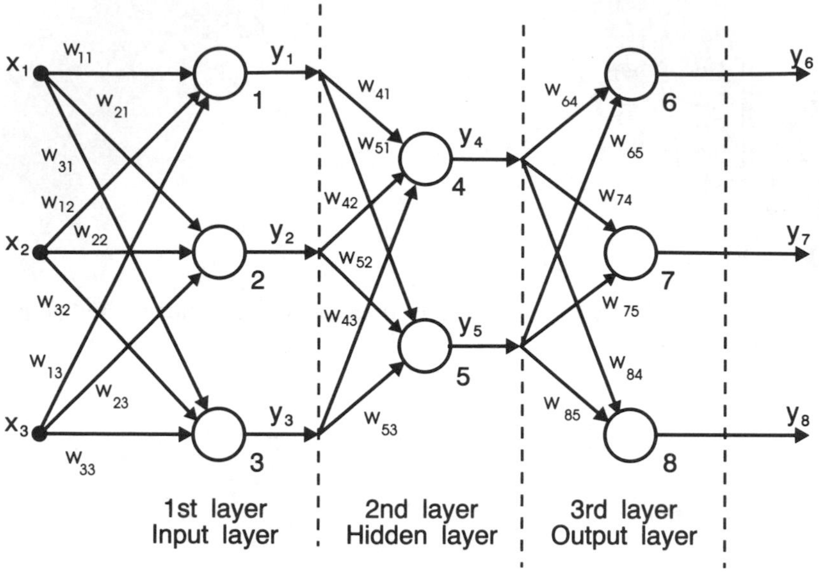

Figure 4.7. Three-layer feed-forward neural network

- Are neural networks a more parsimonious representation of special classes of continuous functions in that they need fewer parameters? If so, what are the characteristics of such functions?
- Given a non-linear mapping $\boldsymbol{h}(\cdot)$ which has to be approximated, what dictates the choice of the number of layers and the number of nodes in each layer of the network?

These questions, which have received considerable attention, have only partial answers at present. However, the most recent results obtained by Albertini and Sontag [1], Albertini *et al.* [2], Sanner and Slotine [31] and Slotine and Sanner [33] are favourable for neural networks.

4.2.3 Feedback (recurrent) networks

The recurrent network, based on the work of Hopfield [11], has been used as a content-addressable memory and in optimization problems. One version of the Hopfield network is shown in Fig. 4.8 and consists of a single-layer neural network in the forward path connected to a delay in the feedback path. The control input in Fig. 4.8 is assumed to be equal to zero. Clearly, the network represents a discrete-time dynamical system with the state vector $\boldsymbol{x}$. The choice of weights determines the equilibrium states of this non-linear dynamical system and thus the specific equilibrium to which the state trajectory converges depends upon the initial conditions $x_1(0), \ldots, x_N(0)$.

Clearly, the following holds:

$$\boldsymbol{y}(t+1) = \boldsymbol{F}(\boldsymbol{W}(t))(\boldsymbol{y}(t)). \tag{4.11}$$

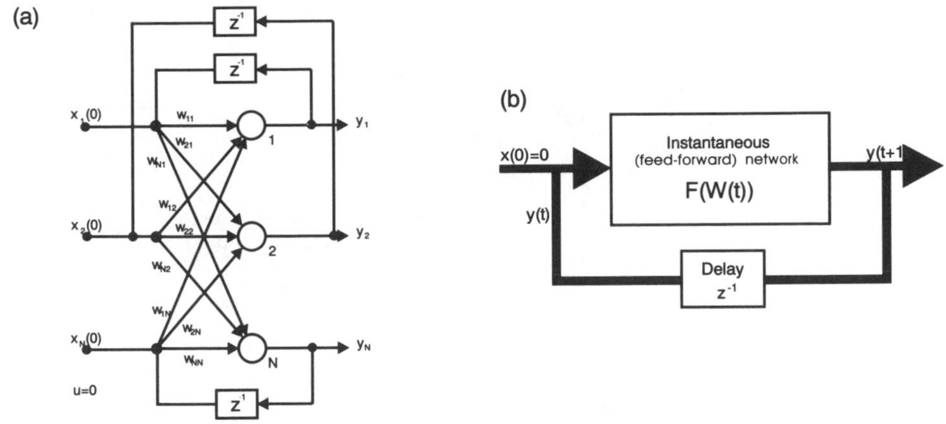

Figure 4.8. Single-layer feedback network. (a) Connection scheme. (b) Block diagram

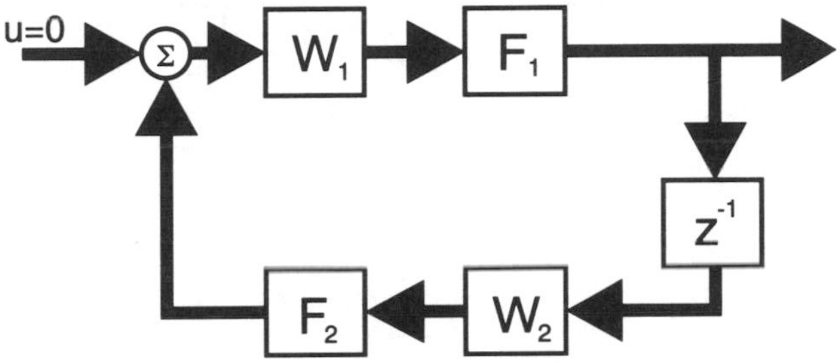

Figure 4.9. Two-layer recurrent network

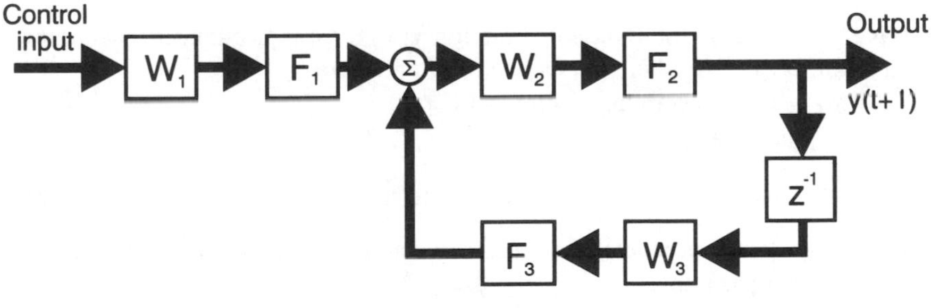

Figure 4.10. Multilayer network in dynamical systems

A two-layer recurrent network is illustrated in Fig. 4.9 again with $\boldsymbol{u} \equiv \mathbf{o}$.

A multilayer recurrent network with control input $\boldsymbol{u}$ and the corresponding feed-forward neural network implementing a feed-forward gain is illustrated in Fig. 4.10. The network represents a non-linear controlled dynamical system.

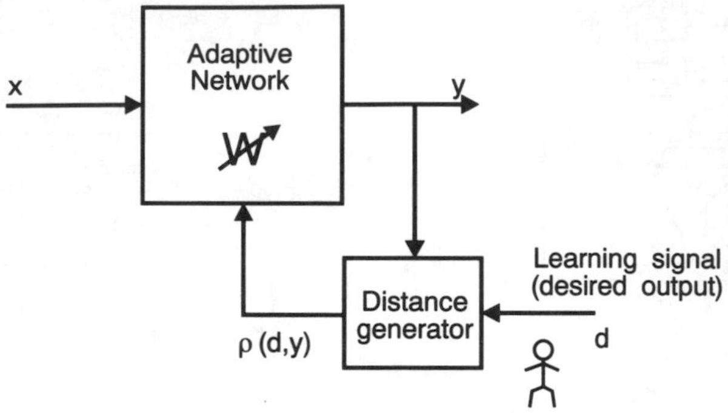

Figure 4.11. Supervised learning

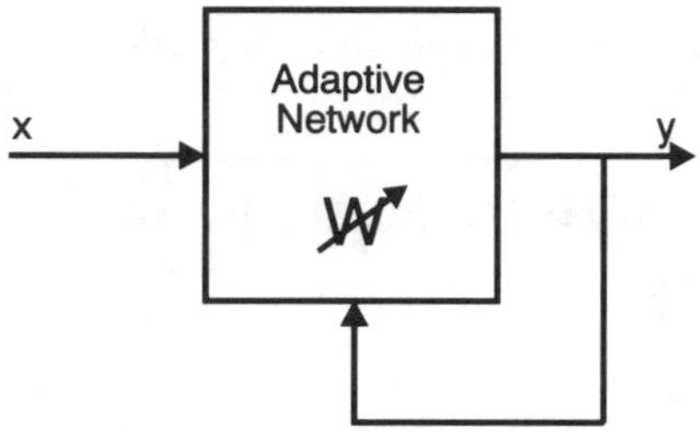

Figure 4.12. Unsupervised learning

4.3 LEARNING IN STATICAL NETWORKS

The classical formulation offered by approximation theory can be expressed as follows:

> Given a continuous multivariable function $\boldsymbol{h}(\boldsymbol{x})$ to be approximated by another function $\boldsymbol{H}(\boldsymbol{x}, \boldsymbol{w})$, where $\boldsymbol{w}$ is a vector of parameters.
> Let $\{\boldsymbol{x}\}$ be a set of training examples of $\boldsymbol{x}$.
> Find $\boldsymbol{w}^*$ such that
>
> $$\rho\left[\boldsymbol{H}(\boldsymbol{x}, \boldsymbol{w}^*), \boldsymbol{h}(\boldsymbol{x})\right] \leq \rho\left[\boldsymbol{H}(\boldsymbol{x}, \boldsymbol{w}), \boldsymbol{h}(\boldsymbol{x})\right] \tag{4.12}$$
>
> for $\boldsymbol{x} \in \{\boldsymbol{x}\}$ and $\boldsymbol{w}$, where $\rho(\cdot, \cdot)$ measures a distance between two functions at the training examples.

Typically, $\rho(\cdot, \cdot)$ is a sum of square differences taken for $\boldsymbol{x} \in \{\boldsymbol{x}\}$. The process of updating the weights in order to minimize ρ and obtain the best value $\boldsymbol{w}^*$ of the parameter $\boldsymbol{w}$ is called learning. The learning may be supervised or unsupervised and these learning structures are explained by Figs 4.11 and 4.12.

4.3.1 Learning rules for a neuron

A general rule is based on work by Amari [3] and is formulated as follows (see Fig. 4.13):

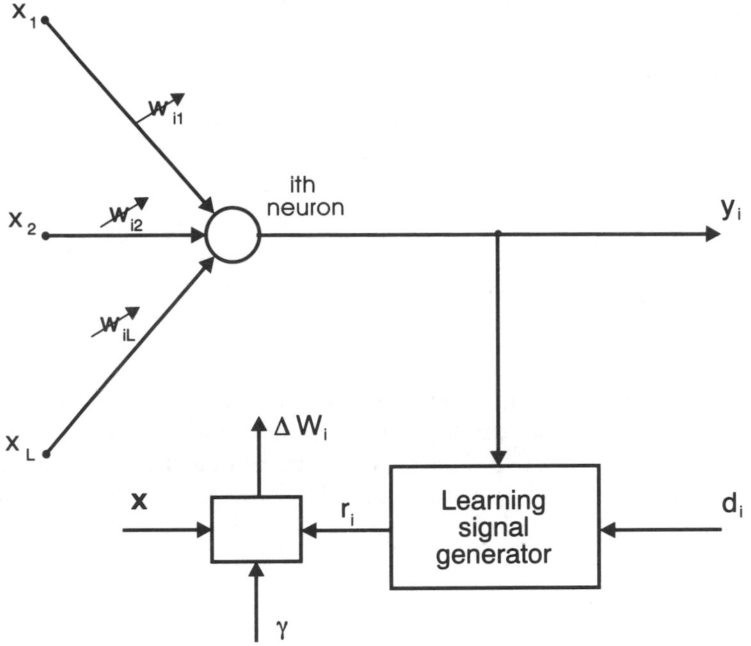

Figure 4.13. Illustration for weight learning rules (provided only for supervised learning)

The weight vector $\boldsymbol{w}_i$ increases in proportion to the product of input $\boldsymbol{x}$ and learning signal r_i.
The learning signal r_i is in general a function of $\boldsymbol{w}_i$, $\boldsymbol{x}$ and sometimes of d_i – the reference, or desired signal. Hence

$$r_i = r_i(\boldsymbol{w}_i, \boldsymbol{x}, d_i). \tag{4.13}$$

The increment of the weight vector is produced as

$$\Delta\boldsymbol{w}_i(t) = \gamma r_i \left[\boldsymbol{w}_i(t), \boldsymbol{x}(t), d_i(t)\right] \boldsymbol{x}(t) \tag{4.14}$$

where γ is a positive number called the learning constant that determines the rate of learning.

The weight vector adopted at time t becomes at the next instant, or learning step,

$$\boldsymbol{w}_i(t+1) = \boldsymbol{w}_i(t) + \gamma r_i \left[\boldsymbol{w}_i(t), \boldsymbol{x}(t), d_i(t)\right] \boldsymbol{x}(t). \tag{4.15}$$

The superscript convention will be also used to index the discrete-time training steps as in the above equation. For the kth step we thus have that

$$\boldsymbol{w}_i^{k+1} = \boldsymbol{w}_i^k + \gamma r_i \left(\boldsymbol{w}_i^k, \boldsymbol{x}^k, d_i^k\right) \boldsymbol{x}^k, \quad k = 1, 2, \ldots \tag{4.16}$$

Continuous-time learning can be expressed as

$$\frac{d\boldsymbol{w}_i(t)}{dt} = \gamma r_i \left[\boldsymbol{w}_i(t), \boldsymbol{x}(t), d_i(t)\right] \boldsymbol{x}(t) \tag{4.17}$$

Learning rules

Hebbian Rule [9]

$$r_i \triangleq f(\boldsymbol{w}_i^T \boldsymbol{x}) = f_i(s_i) \tag{4.18}$$

and

$$\Delta \boldsymbol{w}_i = \gamma f_i(s_i) \boldsymbol{x}. \tag{4.19}$$

Thus, the single weight w_{ij} is adapted using

$$\Delta w_{ij} = \gamma f_i(s_i) x_j. \tag{4.20}$$

This learning requires the weight initialization at small random values around $\boldsymbol{w}_i = \mathbf{o}$ prior to learning. The Hebbian learning rule represents a purely feed-forward, unsupervised learning. The rule states that if the cross product of output and input, or correlation term $f_i(s_i)x_j$ is positive, this results in an increase of weight w_{ij}, otherwise the weight decreases.

Perceptron Learning Rule [28]

For the perceptron rule, the learning signal is the difference between the desired and actual neuron response. Thus, learning is supervised and the learning signal is equal to

$$r_i = d_i - y_i. \tag{4.21}$$

The zero mean threshold activation function (see Fig. 4.2) is used and, therefore,

$$y_i = \text{sgn}(s_i) = \text{sgn}(\boldsymbol{w}_i^T \boldsymbol{x}). \tag{4.22}$$

Weight adjustments in this method are obtained as

$$\Delta \boldsymbol{w}_i = \gamma \left[d_i - \text{sgn}(\boldsymbol{w}_i^T \boldsymbol{x}) \right] \boldsymbol{x} \tag{4.23}$$

and

$$\Delta w_{ij} = \gamma \left[d_i - \text{sgn}(\boldsymbol{w}_i^T \boldsymbol{x}) \right] x_j, \quad \text{for } j = 1, 2, \ldots, L. \tag{4.24}$$

Because the neuron response is only binary, the Eq. (4.24) reduces to

$$\Delta \boldsymbol{w}_i = \pm 2\gamma \boldsymbol{x} \tag{4.25}$$

where a plus sign is applicable if $d_i = 1$, and $\text{sgn}(\boldsymbol{w}_i^T \boldsymbol{x}) = -1$.

Delta Learning Rule [21]

The delta rule is applicable only if an activation function is differentiable and in the supervised mode.

The learning signal for this rule is called "delta" and is defined as

$$r_i = \left[d_i - f_i(\boldsymbol{w}_i^T \boldsymbol{x}) \right] f_i'(\boldsymbol{w}_i^T \boldsymbol{x}). \tag{4.26}$$

This learning rule can be readily derived from the condition of least squared error between y_i and d_i. Calculating the gradient vector with respect to $\boldsymbol{w}_i$ of the squared error defined as

$$E = \frac{1}{2}\left(d_i - y_i\right)^2 \tag{4.27}$$

which is equivalent to

$$E = \frac{1}{2}\left[d_i - f_i(\boldsymbol{w}_i^T \boldsymbol{x})\right]^2 \tag{4.28}$$

we obtain the error gradient vector value

$$\nabla \boldsymbol{E} = -(d_i - y_i) f_i'(\boldsymbol{w}_i^T \boldsymbol{x}) \boldsymbol{x}. \tag{4.29}$$

Since the minimization of the error requires the weight changes to be in the negative gradient direction, we take

$$\Delta w_i = -\eta \nabla \boldsymbol{E} \tag{4.30}$$

where η is a positive constant. We then obtain from Eq. (4.29) and Eq. (4.30)

$$\Delta \boldsymbol{w}_i = \eta (d_i - y_i) f'(s_i) \boldsymbol{x}. \tag{4.31}$$

Widrow–Hoff Learning Rule [36]

The Widrow–Hoff learning rule is applicable to the supervised training of neural networks. It is independent of the activation functions of neurons used since it minimizes the squared error between the desired output value d_i and the neuron's activation value s_i. The learning signal for this rule is defined as

$$r = d_i - s_i = d_i - \boldsymbol{w}_i^T \boldsymbol{x}. \tag{4.32}$$

Thus, the weighting vector increment under this learning rule is

$$\Delta \boldsymbol{w}_i = \gamma (d_i - s_i) \boldsymbol{x}. \tag{4.33}$$

Notice that this rule can be considered as a special case of the delta learning rule if the activation function is simply the identity function, that is if $f_i(s_i) = s_i$.

Speed of convergence and the convergence itself of the learning rule depends on the constant γ. To make the learning algorithm more reliable and efficient its adaptive version was proposed [37] for identity activation function. The constant γ is now updated according to the rule

$$\gamma(\boldsymbol{x}) = \begin{cases} \frac{\boldsymbol{x}'}{\boldsymbol{x}^T \boldsymbol{x}}, & \text{if } \boldsymbol{x}^T \boldsymbol{x} \neq 0, \\ 0, & \text{if } \boldsymbol{x}^T \boldsymbol{x} = 0, \end{cases} \tag{4.34}$$

and the corresponding weight increment is

$$\Delta \boldsymbol{w}_i = \begin{cases} \alpha \gamma(\boldsymbol{x})(d_i - y_i), & \text{if } \boldsymbol{x}^T \boldsymbol{x} \neq 0, \\ 0, & \text{otherwise}, \end{cases} \tag{4.35}$$

where α is constant reduction factor.

We assume that $\boldsymbol{x} \neq 0$ which implies $\boldsymbol{x}^T\boldsymbol{x} \neq 0$. The error dynamics for the adaline whose weights are adjusted by the above rule (which is now the delta rule) can be obtained as follows. Using Eq. (4.34) we have that at the kth iteration

$$E_i^{k+1} \triangleq d_i^{k+1} - y_i^{k+1}$$

and

$$E_i^{k+1} - E_i^k = y_i^k - y_i^{k+1} = -\left[\left(\boldsymbol{w}^{k+1}\right)^T - \left(\boldsymbol{w}^k\right)^T\right]\boldsymbol{x} = -\left(\frac{\alpha E_i^k \boldsymbol{x}^T}{\boldsymbol{x}^T\boldsymbol{x}}\right)\boldsymbol{x} = -\alpha E_i^k .$$

Hence,

$$E_i^{k+1} = (1-\alpha)E_i^k . \tag{4.36}$$

From Eq. (4.36) the error E_i^k converges to 0 at a rate $(1-\alpha)$ if and only if $0 < \alpha < 2$.

A generalization of the above rule to cover case of non-differentiable activation function has been proposed by Sira-Ramirez and Zak [32]. The rule is described as

$$\boldsymbol{w}_i^{k+1} = \begin{cases} \boldsymbol{w}_i^k + \dfrac{\alpha E_i^k \boldsymbol{\theta}(\boldsymbol{x})}{\boldsymbol{x}^T\boldsymbol{\theta}(\boldsymbol{x})}, & \text{if } \boldsymbol{x}^T\boldsymbol{\theta}(\boldsymbol{x}) \neq 0, \\ \boldsymbol{w}_i^k, & \text{otherwise,} \end{cases} \tag{4.37}$$

where $0 < \alpha < 2$.

The error dynamics is described by the Eq. (4.36). The generalized rule dynamics is described by Eq. (4.37) becomes the previous one if the generator $\boldsymbol{\theta}(\cdot)$ is an identity operator, that is if $\boldsymbol{\theta}(\boldsymbol{x}) = \boldsymbol{x}$.

If on the other hand

$$\boldsymbol{\theta}(\boldsymbol{x}) = \begin{bmatrix} sgn(x_1) \\ \vdots \\ sgn(x_2) \end{bmatrix} \tag{4.38}$$

the adaptation algorithms allows non-differentiable activation functions to be considered and convergence is guaranteed.

Applications of this algorithm to adaptive control of unknown non-linear systems are reported in Kuschewski *et al.* [20].

4.3.2 Delta learning rule for multilayer feed-forward networks and back-propagation training algorithm

This section is focused on a training algorithm applied to multilayer feed-forward networks. The algorithm is called the error back-propagation training algorithm, back-propagation for short.

The back-propagation algorithm allows exponential acquisition of input/output mapping knowledge within multilayer networks. Similarly, as in simple cases of delta learning rule training studied before, input patterns are submitted sequentially during back-propagation training. If a pattern is submitted and its classification or association is determined to be erroneous, the current least mean-square classification error is reduced. We shall assume the error to be expressed as

$$E = \frac{1}{2}\sum_{m=1}^{P}(\boldsymbol{d}^m - \boldsymbol{y}(\boldsymbol{x}^m, \boldsymbol{w}))^T(\boldsymbol{d}^m - \boldsymbol{y}(\boldsymbol{x}^m, \boldsymbol{w})) \tag{4.39}$$

where $\boldsymbol{d}^m$ denotes desired output corresponding to the input $\boldsymbol{x}^m$, P is the number of training patterns $(\boldsymbol{d}^m, \boldsymbol{x}^m)$ and $\boldsymbol{y}(\boldsymbol{x}^m, \boldsymbol{w})$ denotes vector output of the network corresponding to the input $\boldsymbol{x}^m$ and weight matrix $\boldsymbol{w}$. Often we take the mean (expected) value of the error E if the training patterns are generated randomly from the training set or if there is a network output measurement error. The delta rule operates then as a stochastic approximation algorithm. We shall simplify further considerations by assuming $P = 1$ and a deterministic case.

During the association or classification phase, the trained neural network itself operates in a feed-forward manner. However, the weight adjustments force the learning rules to propogate exactly forward the input layer.

To derive the back-propagation algorithms we shall consider first the example of the network illustrated in Fig. 4.7. The error can be written as

$$E = \frac{1}{2}[(d_6 - y_6)^2 + (d_7 - y_7)^2 + (d_8 - y_8)^2] \tag{4.40}$$

where d_6, d_7, d_8 denote the network desired outputs corresponding to the prescribed inputs x_1, x_2 and x_3.

The error is, therefore a function of the weights $w_{11}, w_{12}, w_{13}, w_{21}, w_{22}, w_{23}$ (input layer), $w_{64}, w_{65}, w_{74}, w_{84}, w_{85}$ (output layer). By applying gradient descant, the algorithm for adjusting the weights $\boldsymbol{W} = \{w_{ij}\}$ in order to minimize $E(\boldsymbol{W})$ can be written as

$$w_{ij}^{k+1} = w_{ij}^{k} - \gamma\frac{\partial E(w_{ij}^k)}{\partial w_{ij}}, \quad \gamma > 0 \tag{4.41}$$

where k denotes an iteration number.

We shall consider link between neurons 4 and 8 and the corresponding weight w_{84}. The weight w_{84} influences E indirectly according to the following chain:

$$w_{84} \longrightarrow s_8 \longrightarrow y_8 \longrightarrow E \tag{4.42}$$

Therefore, by applying well-known chain rule we obtain that

$$\frac{\partial E(w_{84})}{\partial w_{84}} = \frac{\partial E}{\partial y_8}\frac{\partial y_8}{\partial s_8}\frac{\partial s_8}{\partial w_{84}} = -(d_8 - y_8)f_8'(s_8)y_4 \tag{4.43}$$

let us denote

$$\delta_8(\boldsymbol{x}) \triangleq (d_8 - y_8(\boldsymbol{x}))f_8'(s_8(\boldsymbol{x})) \tag{4.44}$$

and call the $\delta_8(\boldsymbol{x})$ an *equivalent error* associated with the eighth neuron (or output y_8) and corresponding to the input $\boldsymbol{x}$ and weights $\boldsymbol{W}$. Notice that the equivalent error becomes just an error if the activation function $f_8(\ldots)$ is an identity function.

Therefore, the formula for adjusting the weight w_{84} can be written as

$$w_{84}^{k+1} = w_{84}^{k} + \gamma\delta_8 y_4 \tag{4.45}$$

Similarly

$$w_{74}^{k+1} = w_{74}^{k} + \gamma\delta_7 y_4,$$
$$w_{64}^{k+1} = w_{64}^{k} + \gamma\delta_6 y_4,$$
$$w_{85}^{k+1} = w_{85}^{k} + \gamma\delta_8 y_5,$$
$$w_{65}^{k+1} = w_{65}^{k} + \gamma\delta_6 y_5. \tag{4.46}$$

Hence, the output layer adjusted weights as a result of the $(k+1)$th interaction of the training algorithm can be computed by determining the equivalent errors associated with the network outputs and then by using the formulas (4.45) and (4.46).

Let us consider now the hidden layer and the link connecting neurons 1 and 4. The corresponding weight w_{41} is adjusted according to the error gradient descent as

$$w_{41}^{k+1} = w_{41}^{k} - \gamma \frac{\partial E(w_{41}^{k})}{\partial w_{41}} \tag{4.47}$$

The weight w_{41} influences the error E indirectly according to the following chain

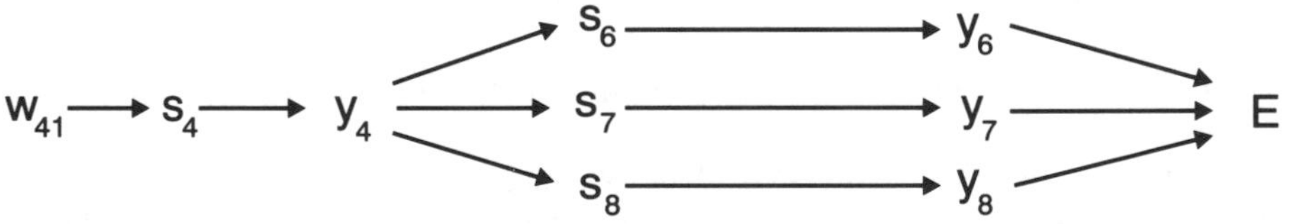

Let us denote a relationship between E and y_4 as $\tilde{E}(y_4)$. Applying the chain rule we obtain

$$\frac{\partial E}{\partial w_{41}} = \frac{\partial \tilde{E}}{\partial y_4}\frac{\partial y_4}{\partial s_4}\frac{\partial s_4}{\partial w_{41}} = \frac{\partial \tilde{E}}{\partial y_4} f_4'(s_4) y_1 \tag{4.48}$$

The derivative $\frac{\partial \tilde{E}}{\partial y_4}$ can be computed as (see the above structure diagram)

$$\begin{aligned}\frac{\partial \tilde{E}}{\partial y_4} &= \frac{\partial E}{\partial y_6}\frac{\partial y_6}{\partial s_6}\frac{\partial s_6}{\partial y_4} + \frac{\partial E}{\partial y_7}\frac{\partial y_7}{\partial s_7}\frac{\partial s_7}{\partial y_4} + \frac{\partial E}{\partial y_8}\frac{\partial y_8}{\partial s_8}\frac{\partial s_8}{\partial y_4} \\ &= -(d_6 - y_6) f_6'(s_6) w_{64} - (d_7 - y_7) f_7'(s_7) w_{74} - (d_8 - y_8) f_8'(s_8) w_{84} \\ &= -\delta_6 w_{64} - \delta_7 w_{74} - \delta_8 w_{84}\end{aligned} \tag{4.49}$$

Let us define an equivalent error for the output of the fourth neuron as

$$\delta_4 \triangleq (\delta_6 w_{64} + \delta_7 w_{74} + \delta_8 w_{84}) f_4'(s_4) \tag{4.50}$$

Then, due to Eqs (4.48), (4.49) and (4.50) the following holds:

$$\frac{\partial E}{\partial w_{41}} = \delta_4 y_1 \tag{4.51}$$

and due to Eq. (4.47)

$$w_{41}^{k+1} = w_{41}^{k} + \gamma\delta_4 y_1 \tag{4.52}$$

Notice that the rule for adjusting the weight in the hidden layer has the same structure as the rule for adjusting the weights in the output layer. However, the equivalent error is now described by a more complicated formula (4.50). We have managed to organize the calculations needed to determine the hidden layer equivalent errors in a recursive way. The procedure, which is the merit of back-propagation algorithms, starts from the output layer equivalent errors and propagates them backwards along the network structure to the considered neuron. The errors are multiplied by the corresponding weights and added. The resulting sum is then multiplied in a standard way by a derivative of the neuron activation function (see Eq. 4.50).

The back-propagation algorithm is an elegant and effective computationally tool for adjusting the network weights.

4.4 LEARNING IN DYNAMICAL NETWORKS

We shall consider in this section dynamical neural networks and also more general dynamical systems containing statical neural networks as their components. The object is to present a suitable training algorithm if a task of the system is to follow a desired trajectory over time period $[t_0, t_f]$. For simplicity we shall concentrate only on discrete-time systems. The error is defined as

$$\boldsymbol{E}(t) \triangleq \boldsymbol{d}(t) - \boldsymbol{y}(t), \quad t = t_0, t_0 + 1, \ldots, t_f \tag{4.53}$$

where $y(t)$ denotes the system output at the instant t.

The performance criterion is defined as

$$J = \frac{1}{2} \sum_{t_0}^{t_f} \boldsymbol{E}^T(t) \boldsymbol{E}(t) \tag{4.54}$$

Clearly, if different input–output training patterns are to be considered the summation in (4.54) must be extended to cover all these patterns as well.

The constant matrix of the weights $\boldsymbol{W}$ is to be found. Again, the gradient descant technique will be employed and a formula for gradient $\nabla J(\boldsymbol{W})$ will be found in a form of the back-propagation algorithm.

4.4.1 Back-propagation through time [30]

We shall investigate a one-layer recurrent network which can be described with the following state equations:

$$\boldsymbol{y}(t+1) = \boldsymbol{\Gamma}(\boldsymbol{y}(t), \boldsymbol{u}(t)) \tag{4.55}$$

where Γ is non-linear mapping and $\boldsymbol{u}$ is control input. Figure 4.14 presents a block diagram of a system which can be described by Eq. (4.55).

The idea of back-propagation through time is to unfold the network through time, i.e. replace the one-layer recurrent network with a feed-forward one with t_f layers (see Fig. 4.15) represented by the same neural network modelling the mapping $\boldsymbol{\Gamma}$.

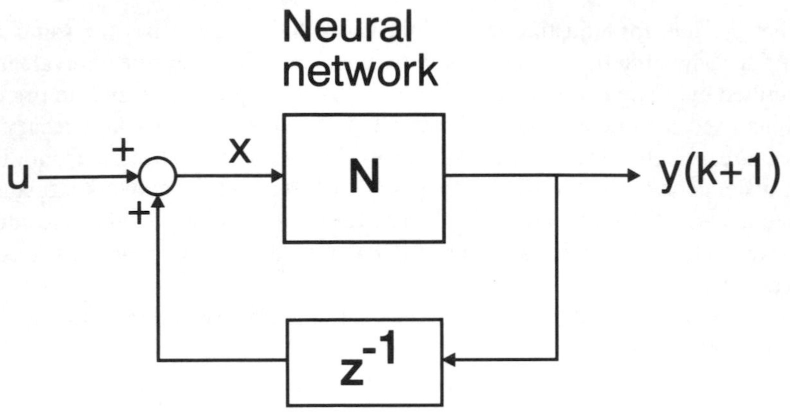

Figure 4.14. N-feed-forward multilayer neural network, z^{-1} – time shift operator

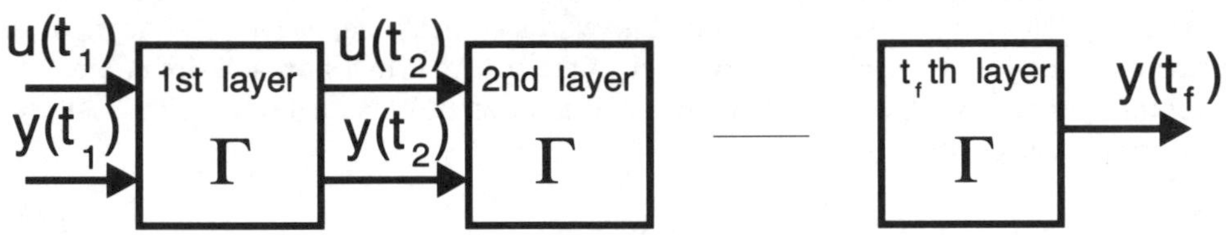

Figure 4.15. Structure of back-propagation through time

It follows from (4.54) that

$$\frac{\partial J(\boldsymbol{W})}{\partial \boldsymbol{w}_i} = \sum_{t_0}^{t_f} \boldsymbol{E}^T(t)\frac{\partial \boldsymbol{E}(t)}{\partial \boldsymbol{w}_i} \tag{4.56}$$

The derivatives $\frac{\partial \boldsymbol{E}(t)}{\partial \boldsymbol{w}_i}$ of the errors at subsequent time instants with respect to the weight $\boldsymbol{w}_i$ can be computed by applying a statical back-propagation scheme at each time instant based on the input produced of the previous time instant and the error corresponding to this time instant.

4.4.2 General dynamical back-propagation

Let us consider the system illustrated in Fig. 4.14 and replace the time shift unit z^{-1} by a general time shift operation represented by transfer function $\boldsymbol{D}(z)$.

Notice, that

$$\frac{\partial \boldsymbol{E}(t)}{\partial \boldsymbol{w}_i} = -\frac{\partial \boldsymbol{y}(t)}{\partial \boldsymbol{w}_i} \tag{4.57}$$

The following holds:

$$\boldsymbol{y}(t+1) = \boldsymbol{N}(\boldsymbol{x}(t+1), \boldsymbol{W}), \qquad \boldsymbol{x}(t+1) = \boldsymbol{u}(t) + \boldsymbol{D}(z)\boldsymbol{y}(k+1) \tag{4.58}$$

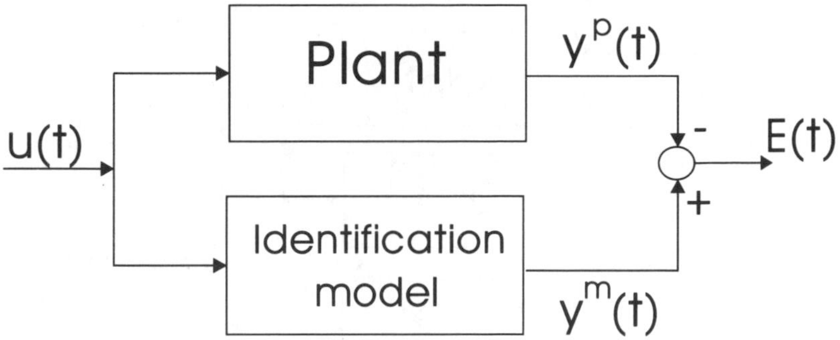

Figure 4.16. Identification

Thus, and

$$\boldsymbol{y}(t+1) = \boldsymbol{N}(\boldsymbol{u}(t) + \boldsymbol{D}(z)\frac{\partial \boldsymbol{y}(t+1)}{\partial \boldsymbol{w}_i}, \boldsymbol{W}) \tag{4.59}$$

Hence,

$$\frac{\partial \boldsymbol{y}(t+1)}{\partial \boldsymbol{w}_i} = \frac{\partial \boldsymbol{N}(\boldsymbol{x}(t+1), \boldsymbol{W})}{\partial \boldsymbol{x}} \boldsymbol{D}(z)\boldsymbol{y}(t+1) + \frac{\partial \boldsymbol{N}(\boldsymbol{x}(t+1), \boldsymbol{W})}{\partial \boldsymbol{w}_i} \tag{4.60}$$

Where arguments of all mappings and functions are taken at the nominal values corresponding to current values of weights under applied control input.

The operator $\boldsymbol{D}(z)\boldsymbol{y}(t+1)$ is comparable to the output components at time instants $t, t-1, t-2, \ldots$. Therefore, Eq. (4.57) constitutes a recursive scheme which can be used to evaluate the error derivatives at subsequent time instant (see Eq. (4.57). The derivatives $\frac{\partial \boldsymbol{N}}{\partial \boldsymbol{x}}$ and $\frac{\partial \boldsymbol{N}}{\partial \boldsymbol{w}_i}$ must be computed separately and evaluated at every time instant.

The above approach can be applied to other dynamical systems including statical neural networks and linear dynamics components.

4.5 IDENTIFICATION

The input and output of a time-invariant, causal discrete-time dynamical plant are $u(\cdot)$ and $y_p(\cdot)$, respectively, where $u(\cdot)$ is a uniformly bounded function of time. The plant is assumed to be stable with a known parameterization but with unknown values of the parameters. The objective is to construct a suitable identification model (Fig. 4.16) which when subjected to the same input u as the plant, produces an output y^m which approximates y^p in a certain sense. We shall choose the same criterion as in Section 4. The model will be used in the form of a neural network.

4.5.1 Forward modelling

The procedure of training a neural network to represent the forward dynamics of a plant will be referred to as forward modelling. The neural network model is placed in parallel

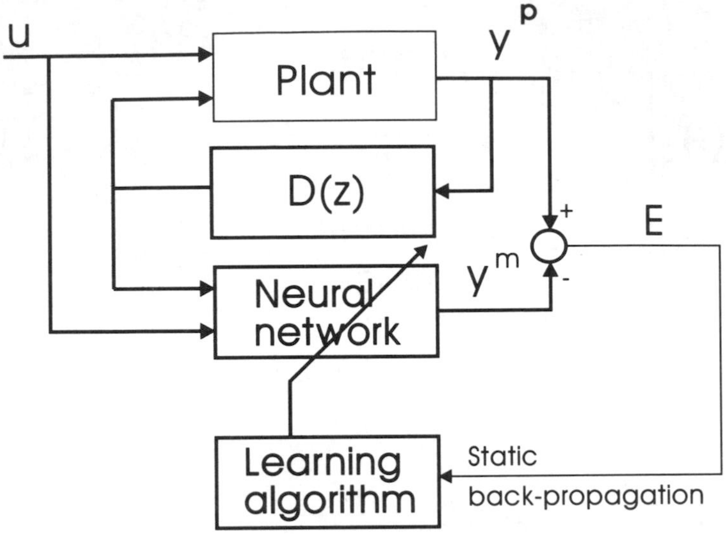

Figure 4.17. Series-parallel identification structure

with system and the error between the system and the network outputs (the prediction error) as the neural network training signal. We shall use a multilayer feed-forward network in order to apply a back-propagation training algorithm.

Let us assume that the plant is governed by the following non-linear discrete-time difference equation:

$$y^p(t+1) = F\left[y^p(t), \ldots, y^p(t-n+1); u(t), \ldots, u(t-m+1)\right] \tag{4.61}$$

Thus, the plant output y^p at time $t+1$ depends on the past n output values and on the past m values of the input u. We concentrate here only on the dynamical part of the plant response; the model does not explicitly represent plant disturbances (for a method of including the disturbance see, e.g., Chen *et al.* [5]). Special cases of the model (4.61) have been considered by Narendra and Parthasarathy [25].

An obvious approach for system modelling is to choose the input–output structure of the neural network to be the same as that of the system. Denoting the output of the network by y^m we then obtain that

$$y^m(t+1) = \hat{F}\left[y^p(t), \ldots, y^p(t-n+1); u(t), \ldots, u(t-m+1)\right] \tag{4.62}$$

In the above, the mapping $\hat{F}(\cdot)$ represents the non-linear input–output map of the network which approximates the plant mapping $F(\cdot)$. Note that the input to the network includes the past values of the plant output but not the past values of the network output (the network has no feedback). The learning statical back-propagation algorithm is used to find optimal values of the network weights. The structure of the model equations (4.62) is called series-parallel. The resulting identification structure is illustrated in Fig. 4.16.

If we assume that after a suitable training period the network gives a good representation of the plant (i.e. $y^m \approx y^p$), then for subsequent post-training purposes the network output

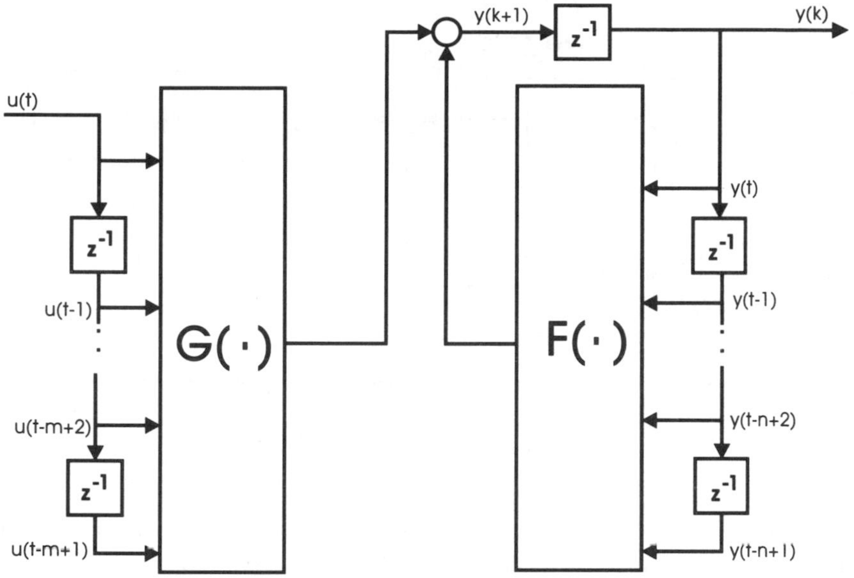

Figure 4.18. Structure of the plant dynamics

itself and its delayed values can be fed-back and used as part of the network input. In this way the network can be used independently of the plant. Such a network is described as

$$y^m(t+1) = \hat{F}\left[y^m(t), \ldots, y^m(t-n+1); u(t), \ldots, u(t-m+1)\right]. \tag{4.63}$$

This structure may also be used from the beginning, that is during the whole process of learning. The structure of (4.63) is called parallel. It may be preferred when dealing with noisy systems since it avoids the problem of bias caused by noise on the plant output. On the other hand the series-parallel scheme (see Fig. 4.17) is supported by stability results. Moreover, the parallel model requires a dynamical back-propagation training algorithm.

Let us consider now a special case of (4.61), namely

$$y^p(t+1) = F\left[y^p(t), \ldots, y^p(t-n+1)\right] + G\left[u(t), \ldots, u(t-m+1)\right] \tag{4.64}$$

Thus, the effects of the input and output values are additive. The structure of (4.64) is illustrated in Fig. 4.18. Clearly, we can apply a general approach presented before. However, it is mostly reasonable to utilize the additive feature of the plant structure. Therefore, the model is described by the following series-parallel equations:

$$y^m(t) = \hat{F}\left[y(t), \ldots, y(t-n+1)\right] + \hat{G}\left[u(t), \ldots, u(t-m+1)\right], \tag{4.65}$$

where the mapping $\hat{F}$ and $\hat{G}$ are implemented by using two separate neural networks. The neural networks weights $\boldsymbol{W}_F$ and $\boldsymbol{W}_G$ are adjusted independently by statical back-propagation algorithm in order to achieve the best approximation of mappings F and G.

The importance of the class of inputs to be used to train learning systems is generally acknowledged. The training set has to be representative of the entire class of inputs that

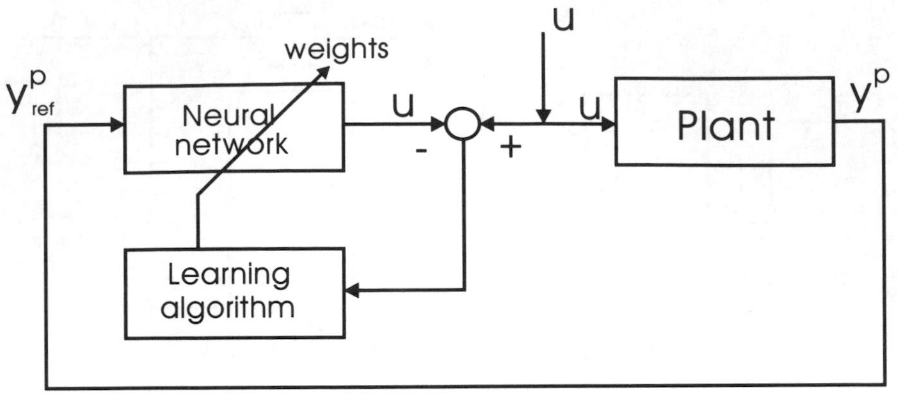

Figure 4.19. Direct inverse modelling

the system may be subject to. This will ensure that the system will respond in the desired fashion even when an input not included in the training set is applied to it. This concept, referred to as persistent excitation, has been extensively treated in conventional control theory both in the context of identification and control problems. The concept of persistent excitation is also found to be important while dealing with the identification and control of non-linear systems using neural networks.

4.5.2 Inverse modelling

The inverse model of a dynamical system yields input for given output. The models play a crucial role in a range of control structures. Some of the structures will be presented in the next section. However, obtaining inverse models raises several important issues. Conceptually the simplest approach is direct inverse modelling as shown in Fig. 4.19. Here, a synthetic training signal (the plant input) is introduced to the system. The plant output is then used as input to the network. The network output is compared with the training signal (the system input) and this error is used to train the network. This structure will clearly force the network to represent the inverse of the plant. However, there are some drawbacks:

- the learning procedure is not “goal directed”; the training signal must be chosen to sample over a wide range of system inputs, and the actual operational inputs may be hard to define *a priori*. The actual goal in the control context is to make the system output behave in a desired way, and thus the training signal in direct inverse modelling does not correspond to the explicit goal;

- if the non-linear system is not one-to-one, then an incorrect inverse can be produced.

The first point is strongly related with the general concept of persistent excitation.

A second approach to inverse modelling which aims to overcome these problems is known as specialized inverse learning [27]. The specialized inverse learning structure is shown in Fig. 4.20. In this approach the network inverse model precedes the system and

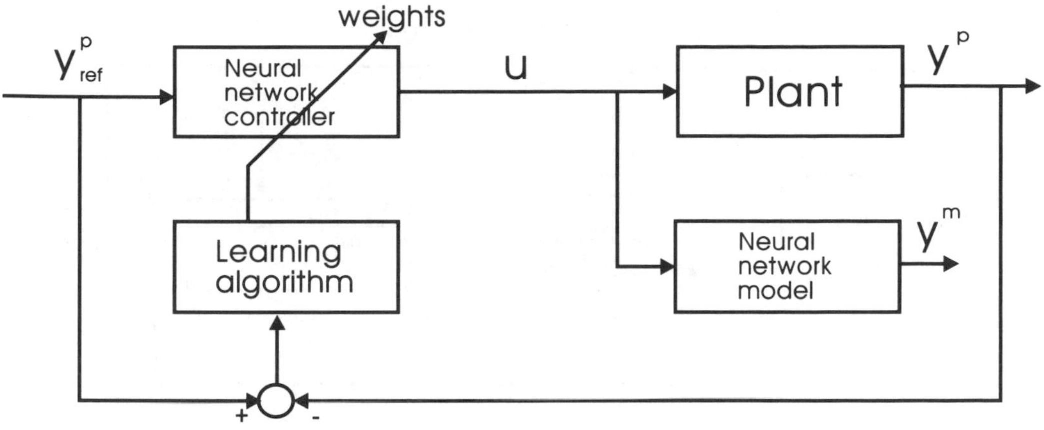

Figure 4.20. Specialized inverse modelling

receives as input a training signal which spans the desired operational output space of the controlled system (i.e. it corresponds to the system reference signal). This learning structure also contains a trained forward model of the system (e.g. a network trained as described in the Section 5.1) placed in parallel with the plant. The error signal for the training algorithm in this case is the difference between the training signal and the system output (it may also be the difference between the training signal and the forward model output if the system is noisy). We can show that using the plant output we can produce an exact inverse even when the forward model is not exact; this is not the case when the forward model output is used. The error may then be propagated back through the forward model and then the inverse model; only the inverse network model weights are adjusted during this procedure. Thus, the procedure is effective at learning and identity mapping across the inverse model and the forward mode; the inverse model is learned as a side effect. In comparison with direct inverse modelling, the specialized inverse learning approach possesses the following features:

- The procedure is goal directed since it is based on the error between desired system outputs and actual outputs. In other words, the system receives inputs during training which correspond to the actual operational inputs it will subsequently receive.

- In case in which the system forward mapping is not one-to-one a particular inverse (pseudo-inverse) will be found. The problem of bias can also be handled.

Let us now consider the input–output structure of network modelling the system inverse. From Eq. (4.61) the inverse F^{-1} leading to the generation of $u(t)$ would require knowledge of the future value $y^p(t+1)$. To overcome this problem we replace this future value with the value $y^p_{ref}(t+1)$ which we assume is available at time t. This seems to be a reasonable assumption since y^p_{ref} is typically related to the reference signal which is normally known one step ahead. Thus, the non-linear input–output mapping relation of the

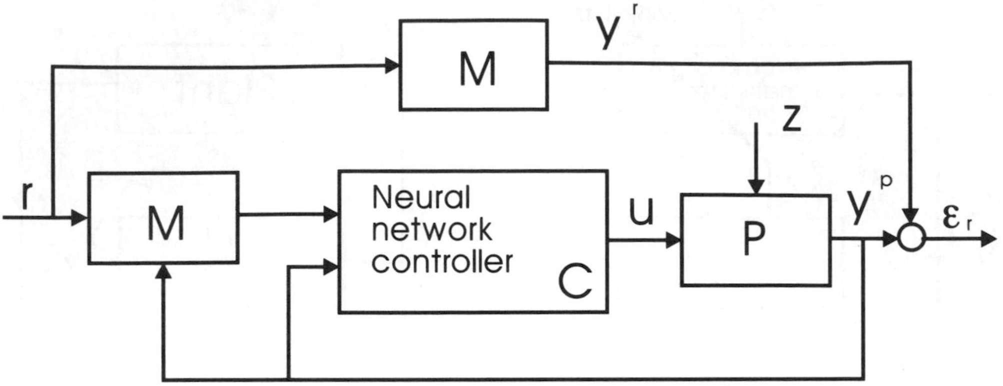

Figure 4.21. Model reference structure

network modelling the plant inverse is

$$u(t) = F^{-1}\left[y^p(t), \ldots, y^p(t-n+1); y^p_{ref}(t+1); u(t-1), \ldots, u(t-m+1)\right] \quad (4.66)$$

that is the inverse model network receives as inputs the current and past system outputs, the training (reference) signal, and the past values of the system outputs. Where it is desirable to train the inverse without the plant (as discussed before) the values of y^p in the above relation are simply replaced by the forward model outputs y^m.

4.6 CONTROL

Models of dynamical systems and their inverses have immediate utility for control. In the control literature a number of well-established and deeply analysed structures for the control of non-linear systems exist; we shall focus on those structures that have a direct reliance on system forward and inverse models. We assume that such models are available in the form of neural networks which have been trained using the techniques outlined above. It is beyond the scope of this work to provide a full survey of all neural network based architecture available in the literature. We shall focus instead on presenting key directions and descriptions of the representative control structures.

4.6.1 Model reference control [25]

The desired performance of the closed-loop system is specified through a stable reference model M, which is defined by its input–output pair $\{r(t), y^r(t)\}$. The control system attempts to make the plant output $y^p(t)$ mach the reference model output asymptotically, i.e. $\lim_{t\to\infty} \|y^r(t) - y^p(t)\| \leq \varepsilon$ for some specified constant $\varepsilon \geq 0$. The model reference control structure for non-linear systems utilizing the connectionist model is illustrated in Fig. 4.21. In this structure the error defined above is used to train the network acting as the controller. Clearly, this approach is related to training of an inverse plant model as presented in the previous section.

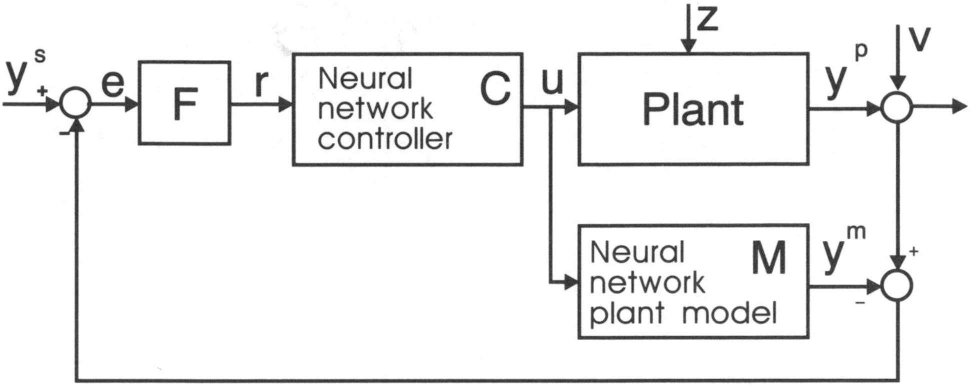

Figure 4.22. Internal model control structure

In general, the training procedure will force the controller to be a "detuned" inverse, in a sense defined by the reference model. The overall scheme can be viewed as direct adaptive control.

4.6.2 Internal model control (IMC)

In IMC the role of system forward and inverse models is emphasized [7]. In this structure, system forward and inverse models are used directly as elements within the feedback loop. IMC has been thoroughly examined with the application of standard robustness and stability analysis. Moreover, IMC extends reality to non-linear systems control. A system model is placed in parallel with the real system. The difference between the system and model outputs is used for feedback purposes. This feedback signal is then processed by a controller subsystem in the forward path; the properties of IMC dictate that this part of the controller should be related to the system inverse. Given a network model for the system forward and inverse dynamics, the realization of IMC using neural networks is straightforward [13]. The system model M and the controller C (the inverse model) are realized using the neural networks as illustrated in Fig. 4.22. The subsystem F is typically a linear filter which is designed to provide the desirable robustness and tracking response of the closed-loop system. We should point out that the applicability of IMC is limited to open-loop stable systems.

4.6.3 Predictive control

In the realm of optimal and predictive control methods the receding horizon technique has been introduced as a natural, computationally feasible feedback law. It has been proven that the method has the desired stability properties for non-linear systems.

In this approach a neural network model provides a prediction of a plant's future response over a specified horizon. The predictions supplied by the network as passed to a numerical optimization routine which attempts to minimize a specified performance criterion in the calculation of a suitable control signal.

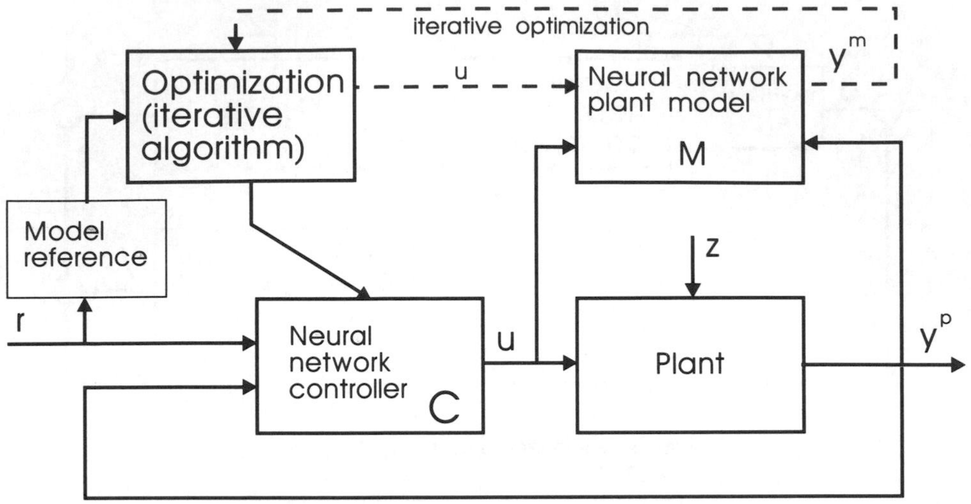

Figure 4.23. Model predictive control structure

The control signal $\tilde{u}$ is chosen to minimize the quadratic performance criterion which compromises between the tracking error and the control cost

$$J = \sum_{j=N_1}^{N_2} (y^r(t+j) - y^m(t+j))^2 + \sum_{j=1}^{N_2} \lambda_j (\tilde{u}(t+j-1) - \tilde{u}(t+j-2))^2 \quad (4.67)$$

subject to the constraints and subject to the equality constraints introduced by the dynamical model itself.

Here, the constants N_1 and N_2 define the horizons over which the tracking error and control increments are considered. The values of λ are the control weights.

Once the iterative optimization algorithm finds the optimal solution u it is applied to the plant. The actual value of the plant output y^p is measured and jointly with the reference signal r and y^p is sent to another neural network. This network is a controller and is trained to produce the same control output u as the optimization routine. As the result the non-linear-feedback control law is obtained. An overall control structure is presented in Fig. 4.23.

4.6.4 Self-learning controller [26]

We shall consider a discrete-time dynamical system

$$y(t+1) = F[y(t), u(t)] \quad (4.68)$$

where the function F is non-linear and unknown.

A control problem is to provide the correct input vector $\{u(t)\}$ to drive the plant from an initial state to a subsequent desired state y^d. The objective is to design the state-feedback controller. We shall use a neural network architecture for the controller and derive a suitable

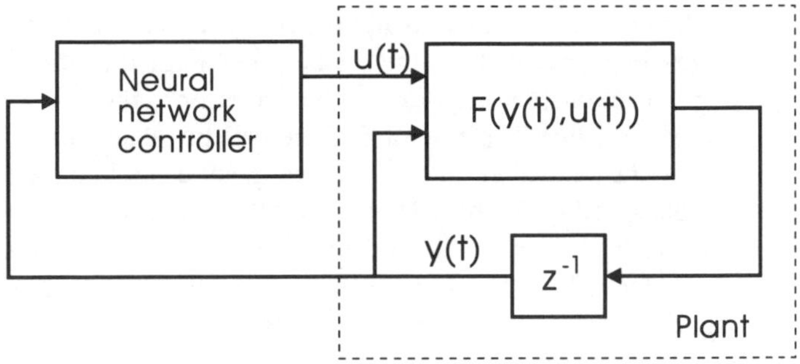

Figure 4.24. Control structure for the self-learning controller

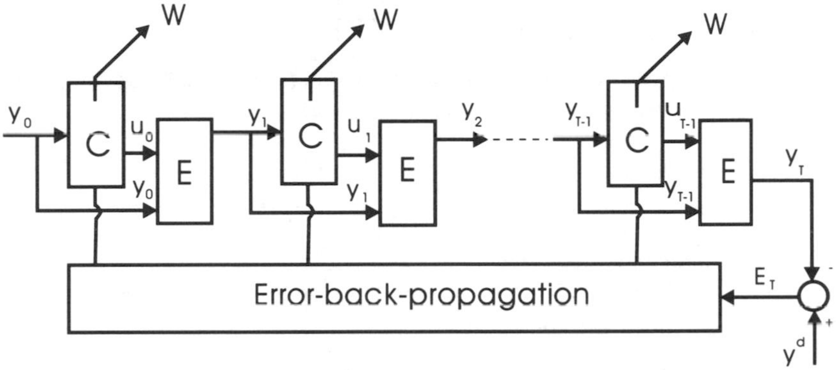

Figure 4.25. Training with back-propagation. C – controller, E – plant emulator

learning algorithm to adjust its weights. The resulting controller is non-linear adaptive. The control system structure is illustrated in Fig. 4.24.

The learning algorithm requires the plant emulator which can be found in the form of a neural network by using the techniques presented in Section 5. The training procedure is as follows. The controller learns to drive the plant emulator from an initial state y_0 to the desired state y^d in T time steps. Learning takes place during many trials or runs, each starting from an initial state and terminating at a final state y_T. The objective of the learning process is to find a set of controller weights that minimizes the error function which is an average of $||y^d - y_K||^2$ over the set of initial states y_0. The training process is illustrated in Fig. 4.25. The training starts with the neural net plant emulator set in a random initial state y_0. Because the neural net controller initially is untrained, it will produce an erroneous control signal u_0, to the plant emulator and plant itself. The plant emulator will then move to the next state $y_1 = y(1)$, and this process continues for T time steps. At this point the plant is at the state $y_T = y(T)$. The number of time steps T should be determined by the designer. We would now like to modify the weights in the controller network so that the square error $||y^d - y_T||^2$ will be less at the end of the next run. To train the controller, we need to know the error in the controller output u_k for each time step k. Unfortunately, only the error in the final plant state, $y^d - y_T$ is available. However, because the plant emulator is a neural network, we can propagate back the final error through the plant emulator using standard back-propagation algorithm to get an equivalent error in the Tth stage. This error can be used to train the controller. The emulator, therefore, translates the error in the final plant state to the error in the controller output. The real plant cannot be used here because the error cannot be propagated through it. This is why the neural net emulator is needed. the error continues to be back-propagation scheme, and the controller's weight change is computed for each stage. The weight changes from all the stages obtained from the back-propagation algorithm are added together and then added to the controller's weights. This completes the training for one run.

The overall scheme can be viewed as a self-learning controller. However, the controller parameter adjustment algorithm performs off-line and does not use the data from real plant.

4.6.5 Non-linear self-tuning adaptive control [5]

We shall consider a class of single-input, single-output (SISO) feedback linearizable systems described by the discrete-time input–output equation:

$$\begin{aligned} y_{t+1} &= F(y_t, y_{t-1}, \ldots, y_{t-p}, u_{t-1}, \ldots, u_{t-p}) \\ &\quad + G(y_t, y_{t-1}, \ldots, y_{t-p}, u_{t-1}, \ldots, u_{t-p})u_t \end{aligned} \tag{4.69}$$

The functions F and G are unknown.

If we knew both F and G of (4.69), we could use the following control, and the system would exactly track the desired output y^d_{t+1}:

$$u_t = -\frac{F(\cdot)}{G(\cdot)} + \frac{y^d_{t+1}}{G(\cdot)} \tag{4.70}$$

Since $F(\cdot)$ and $G(\cdot)$ are unknown, neural networks can be used to "learn" to approximate these functions and generate suitable controls. In order to simplify the problem and focus on the control mechanism, discussion will be limited to the first-order system.

$$y_{t+1} = F(y_t) + G(y_t)u_t \tag{4.71}$$

Although the function $G(\cdot)$ is not known, we can assume that the sign of $G(y_t)$ is known along plant trajectories. The system (plant) can be modelled by the neural network illustrated in Fig. 4.26, where $\boldsymbol{W} = (w_0, w_1, \ldots, w_{2p})$ and $\boldsymbol{V} = (v_0, v_1, \ldots, v_{2q})$ are the weights of the neural networks approximating the functions $F(\cdot)$ and $G(\cdot)$, respectively (see Fig. 4.26).

Therefore, the plant neural network model is described as

$$y^m_{t+1} = F^m[y_t, \boldsymbol{W}(t)] + G^m[y_t, \boldsymbol{V}(t)]u_t \tag{4.72}$$

Notice that the model equation is series-parallel (see Section 5). Notice also that $F^m(0, \boldsymbol{W}) = w_0$ and $G^m(0, \boldsymbol{V}) = v_0$. The linear neurons are labelled "L". They are able to scale and shift incoming signals. The neurons labelled "H" are non-linear. We assume that there are enough neurons to approximate sufficiently accurately (with desired accuracy) the unknown functions $F(\cdot)$ and $G(\cdot)$. Although there exist in the literature general results

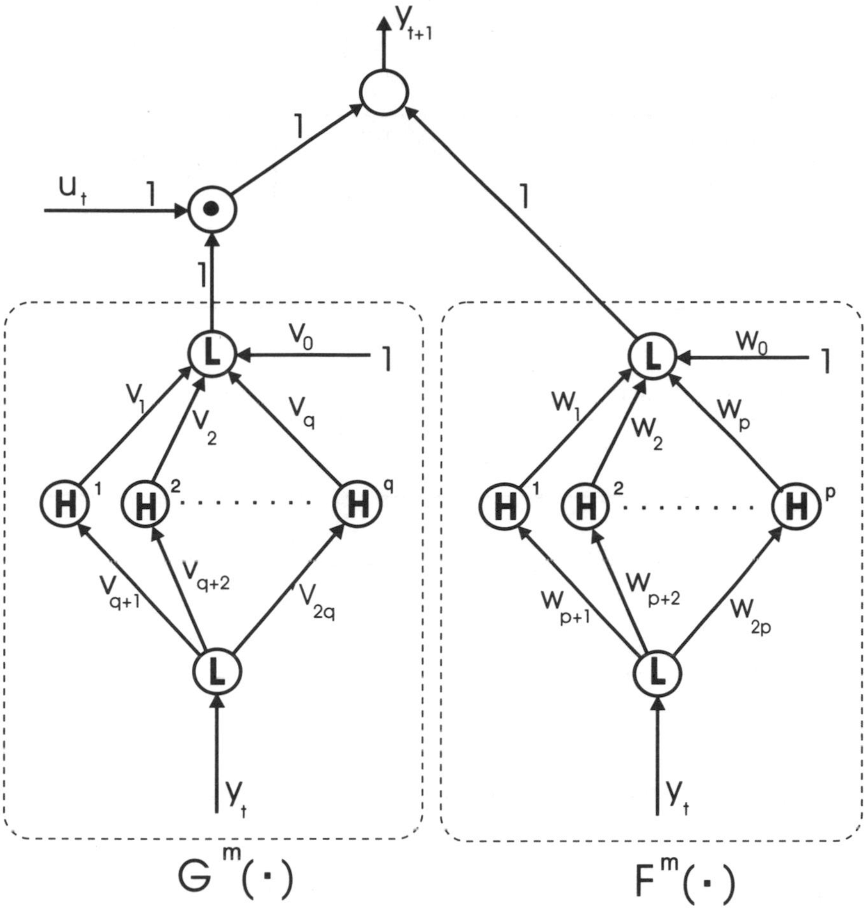

Figure 4.26. The neural network model

relating the neuron number and the accuracy of approximation, in practice this number must be found by trial and error.

The control task is for the plant output to track the command $\{y_d(t)\}$. The overall control system is illustrated in Fig. 4.27. At time-step t, the following control (see Eq. 4.70) is applied to the plant (4.71):

$$u_t = -\frac{F^m[y_t, \boldsymbol{W}(t)]}{G^m[y_t, \boldsymbol{V}(t)]} + \frac{y_{t+1}^d}{G^m[y_t, \boldsymbol{V}(t)]} \tag{4.73}$$

In contrast to "open-loop" training, the feedback dramatically changes the role of weights of the neural network. The output of the plant depends on the weights of the neural net and serves as the desired output of the neural network:

$$y_{t+1} = F(y_t) + G(y_t)\left[-\frac{F^m[y_t, \boldsymbol{W}(t)]}{G^m[y_t, \boldsymbol{V}(t)]} + \frac{y_{t+1}^d}{G^m[y_t, \boldsymbol{V}(t)]}\right] \tag{4.74}$$

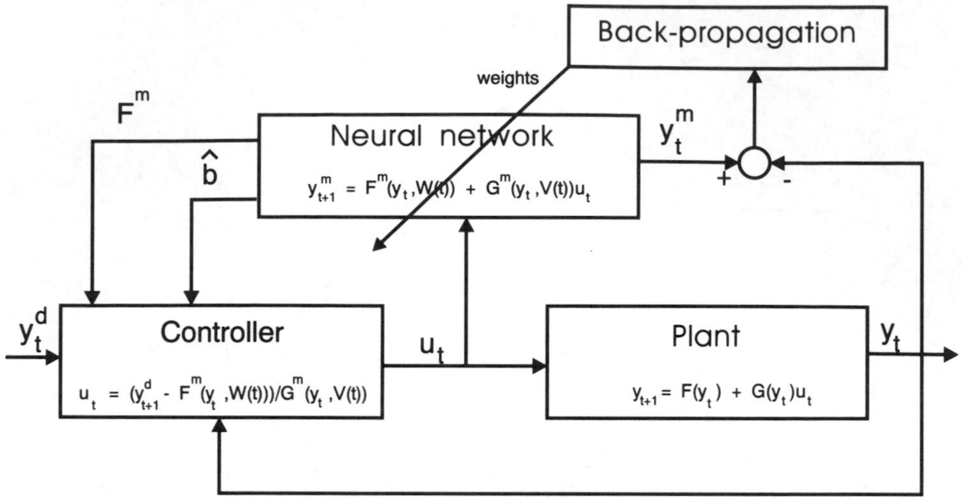

Figure 4.27. Neural network self-tuning control system

If (4.73) is substituted into (4.72), the output of the neural network is y_t^d which is independent of $\boldsymbol{W}(t)$ and $\boldsymbol{V}(t)$. Let us define the output error as

$$E(t) = \frac{1}{2}(y_{t+1}^d - y_{t+1})^2 \tag{4.75}$$

Then, $\boldsymbol{W}(t)$ and $\boldsymbol{V}(t)$ are to be adjusted such that $E(t)$ can be reduced. We can easily verify that

$$\frac{\partial E(t)}{\partial w_i(t)} = \frac{G(y_t)}{G^m[y_t, \boldsymbol{V}(t)]}\left[\frac{\partial F^m[y_t, \boldsymbol{W}(t)]}{\partial w_i(t)}\right] E(t+1) \tag{4.76}$$

and

$$\frac{\partial E(t)}{\partial v_i(t)} = \frac{G(y_t)}{G^m[y_t, \boldsymbol{V}(t)]}\left[\frac{\partial G^m[y_t, \boldsymbol{V}(t)]}{\partial v_i(t)}\right] u_t E(t+1) \tag{4.77}$$

The quantities

$$\frac{\partial F^m[y_t, \boldsymbol{W}(t)]}{\partial w_i(t)} \quad \text{and} \quad \frac{\partial G^m[y_t, \boldsymbol{V}(t)]}{\partial v_i(t)} \tag{4.78}$$

can be calculated by employing the standard back-propagation algorithm.

The quantity $G(y_t)$ is not known, but its sign is assumed to be known. Therefore, $G(y_t)$ is replaced by $\text{sgn}[G(y_t)]$ before (4.76) and (4.77) are used in the following gradient descant updating rules:

$$w_i(t+1) = w_i(t) - \gamma_t \frac{\text{sgn}[G(y_t)]}{G^m[y_t, \boldsymbol{V}(t)]}\left[\frac{\partial F^m[y_t, \boldsymbol{W}(t)]}{\partial w_i(t)}\right] E(t+1) \tag{4.79}$$

$$v_i(t+1) = v_i(t) - \mu_t \frac{\text{sgn}[G(y_t)]}{G^m[y_t, \boldsymbol{V}(t)]}\left[\frac{\partial G^m[y_t, \boldsymbol{V}(t)]}{\partial v_i(t)}\right] u_t E(t+1) \tag{4.80}$$

The positive scalars γ_t, μ_t specify the learning rates at time-step t. With y_{t+1}, $\boldsymbol{W}(t+1)$ and $\boldsymbol{V}(t+1)$ available (see 4.74, 4.79 and 4.80) the functions values $F^m[y_{t+1}, \boldsymbol{W}(t+1)]$ and $G^m[y_{t+1}, \boldsymbol{V}(t+1)]$ are calculated by the neural networks. Then $F^m[y_{t+1}, \boldsymbol{W}(t+1)]$, $G^m[y_{t+1}, \boldsymbol{V}(t+1)]$ and y^d_{t+2} can be used to generate u_{t+1} according to (4.73) for the next step.

The learning rates can be computed on-line based on current values of the plant output so that it can be shown by using Lyapunov theory that with these rates the tracking error converges to zero. It is possible to extend the scheme to more general situation.

The controller derived here uses on-line data from the plant to adjust its parameters and generate the control signal. It is a truly self-tuning non-linear controller. The controller has been recently applied to control of stator magnetizing current in an AC drive system and performed much better than classical adaptive schemes which use a linear parameterized model of a non-linear plant [19].

4.7 CONCLUSIONS

Neural networks have a history of over fifty years but have found solid application only in the last decade. They are able to perform complex functions in various fields including pattern recognition, classification, identification and modelling, speech, vision and control systems. The work presented has concerned mostly identification and control problems. The network architecture has first been presented, including statical and dynamical, single and multilayer, and feedback (recurrent) networks. One of the most difficult problems, and still of great interest to scientists, is learning. Beginning from the very basic *Hebbian rule* the most common techniques are: *perceptron learning rule, Delta learning*, and the *Widrow–Hoff rule*. Special attention has been paid to the most efficient learning algorithm for multilayer networks, namely *back-propagation*. Learning in dynamical networks has also been presented.

Learning algorithms are the basis for an introduction to identification problems. Forward and inverse modelling have been discussed and proper structure of identification systems has been presented. Neural models obtained via identification are intended to become a part of the control structure. *Model reference control* has been presented; the neural inverse model has been used. Neural modelling of the controller as well as the plant is a property of *internal model control*. Models provided by neural networks are non-parametric. One of the most useful properties of ANNs is prediction ability. *Predictive control* is the structure in which this ability is used in a special way. The *self-learning controller* proposed by Nguyen and Widrow has also been presented. One of the most interesting structures of a neural controller has been proposed lately by Chen: *non-linear self-tuning adaptive control*. It is especially promising for control of more complex non-linear plants.

Artificial neural networks are a very hot area and we can expect many exciting results in the very near future. The work presented here is a short but complete (with respect to identification and control problems) overview of the state of the art.

4.8 LABORATORY EXERCISES

4.8.1 MATLAB software tools applied in laboratory course 4

The Neural Network Toolbox provides the basic help in creating software involving neural nets in identification and control. The ANN area covers a large number of problems; therefore the NN Toolbox covers such topics as pattern recognition and classification, pattern association and associative memory, and self-organizing. Identification and control problems have no special functions in the toolbox. However, the functions provided are mostly sufficient to solve basic problems. Details of the toolbox can be found in the user's guide. Here, only comments are given.

The transfer function of the neuron can be chosen as sigmoidal, hard-limit, linear or saturated linear. The latest version of the NN Toolbox (version 2.0b, released 6 September 1994) provides radial functions and the Levenberg–Marquardt learning algorithm. These two seem to be most important extensions of the toolbox for identification and control.

The structure of the network can be single, double or triple layered. If more layers are to be incorporated, the proper m-files should be changed by the user. Such a situation is common in back-propagation through time. In principle, the network to be simulated is statical multi-layer feed-forward (see Section 4.2.2). Recurrent networks (Section 4.2.3) can be easy simulated by introducing feedback. The examples in Sections 4.8.3 and 4.8.4 illustrate recurrent networks.

According to Section 4.3.1, all learning rules are described in the toolbox. Clearly, the back-propagation algorithm is also included. Despite its basic form (see Section 4.3.2) certain modifications of the back-propagation algorithm are provided. The modifications, namely the momentum method, improvement of initial conditions and adaptive learning rate, are discussed in the examples of Section 4.8.2. The examples provided could be easily customized to include the Levenberg–Marquardt learning rule as well as to use radial basis functions.

Off-line identification (Section 4.8.3) of the neural net models is easy and efficient using the NN Toolbox, because all data constituting input and output plant measurements can be collected in matrices of so-called patterns. Learning is thus processing of the patterns, which is done using simple functions. Certain additional functions help in observation and evaluation of the progress of the learning. However, we should remember that for a large problem the convergence of learning could be too slow due to the way that MATLAB interprets its commands.

On-line (recursive) identification (Section 4.8.4) is not so easy in the NN Toolbox. The user should build his/her own function. The facilities of SIMULINK can be used here as in the examples provided in Section 4.8.4.

Other, more detailed comments concerning software tools will be given in the following sections.

4.8.2 Single neuron and learning

Perfect approximation

Example 1 – the choice of initial point

Sigmoidal function of tangent type can be defined as

$$s(x) = \frac{1}{1 + e^{-(Wx+B)}} \tag{4.81}$$

There are two parameters of the function, namely W (weight) and B (bias), which have to be chosen. Clearly, if the function has to approximate two points, we can calculate the values of the parameters exactly (perfect approximation).

Let

$$x_1 = -2 \qquad s_1 = 0.2$$

$$x_2 = 3 \qquad s_1 = 0.7$$

From the set of equations

$$s_1 = \frac{1}{1 + e^{-(Wx_1+B)}} \tag{4.82}$$

$$s_2 = \frac{1}{1 + e^{-(Wx_2+B)}} \tag{4.83}$$

one can obtain

$$W = \frac{1}{x_1 - x_2} ln \left(\frac{s_1}{s_2} \frac{1-s_2}{1-s_1} \right) \tag{4.84}$$

$$B = ln \left(\frac{s_1}{1-s_1} \right) - Wx_1$$

In the above example

$$W = 0.4467$$

$$B = -0.4929$$

Use the MATLAB program `nn1_1` to check how perfectly the approximation works. Compute the exact values of W and B and compare them with the values obtained in the learning process.

Quick help

1. Enter the program `nn1_1` at the MATLAB prompt.
2. According to the display at the command window, enter two points to be approximated by the sigmoidal function. This can be done by clicking with the mouse at the desired place in the graph window. The entry is finished after clicking out of the figure at the graph window.

3. The program responds with exact values inputted. **P** refers to x (**p**atterns), **T** refers to s (**t**arget). The error surface is calculated and plotted. The plot shows how error depends on W and B.

4. Using formulas (4.82) and (4.83) exact values of parameters W and B should be calculated. Note where they are placed on the error graph.

5. Press any key in the command window to obtain a contour plot of the error surface. According to the display in the command window, choose the starting point of the learning procedure by clicking with the mouse at the desired point.

6. The program responds with default values of the learning parameters. Do not change them for the time being. On the contour plot the progress of learning can be seen. Repeat the learning by choosing different initial points. This can provide the answers for question 1 below.

7. By clicking outside of the contour plot, the network error versus epochs is plotted for the latest learning process. It can happen that for a proper initial point choice, the number of epochs is less than the maximum number, showing very efficient learning.

8. The next plot, activated by pressing any key in the command window, shows the efficiency of approximation. Because the maximum number of epochs is not very high and the error-goal is not very small, the approximation cannot be perfect. This can be seen om the next plot, showing two bars proportional to the errors for the corresponding points to be approximated.

Questions

1. The number of epochs and learning rate are constant. How does the choice of initial point influence the progress of the learning process?

2. How does the speed of learning depend on the choice of initial point?

3. How far is the final point from the exact one?

Example 2 – parameters of learning

The approximation can be made more accurate if the parameters of the learning procedure are chosen properly. Repeat the above example changing the parameters.

Quick help

1. Again use the program `nn1_1`.

2. Begin with the choice of number of epochs. If the final results are not satisfactory but the maximum number of epochs is not reached, make the error-goal smaller.

3. Try to make the learning process faster by making the learning rate greater.

Questions

1. How do the number of epochs and error-goal influence the approximation accuracy and time needed for learning?

2. What is the effect of making the learning rate too large or too small?

Imperfect approximation

Example 1 – local minima

If the number of points is greater than two, perfect approximation is no longer possible. The error surface become more corrugated and some local minima arise. The learning process becomes more difficult: a wrong choice of initial point can result in a solution at a non-global minimum.

Using the program `nn1_1`, check the possibility of approximating the set of points with a sigmoidal function. Use different points to be approximated.

Quick help

1. Enter 4–6 points constituting a shape similar to the sigmoidal function and observe the error surface as well as the final approximation.

2. Enter 4–6 points far from the shape of the sigmoidal function and observe the local minima. Try to find initial points which provide best and worse results of learning. Observe the speed of learning.

3. Repeat the last experiments, changing the number of epochs.

Questions

1. Explain the mechanism of local minima formation. Try to interpret it in geometrical terms.

2. How does the number of epochs influence the learning progress?

3. How does the network error depend on initial conditions and on the shape of the error surface?

4. Interpret the output vector errors (bar plot).

Example 2 – influence of learning rate

Example 2 shows that the basic problem is the choice of learning rate. Repeat the experiments from example 1 with different values of learning rate. Use a set of points to be approximated which gives rise to a corrugated error surface. Observe the behaviour of the error trajectory during learning.

Questions

1. Is it possible to make the speed of learning arbitrarily high providing the initial points are chosen properly?

2. Give an idea of improving the learning process. **Hint:** interpret the learning process as a ball rolling over a surface of varying friction; what is the criterion of the quality of the learning progress?

Improvement of learning

The previous section showed that the learning strategy plays a crucial role in determining the accuracy of approximation. In the following, some improvements of the learning procedure will be tested.

Example 1 – momentum

Use the program `nn1_2` to learn how the momentum method improves the behaviour of the error in the neighbourhood of the minimum.

Quick help

1. Enter the program `nn1_2` and input the set of 4–6 points to be approximated. The set should be chosen in such a way that some corrugation of the error surface is obtained.

2. Choose the standard learning at the beginning; the initial point should not be chosen very close to the minimum.

3. Try to increase the learning rate in order to obtain oscillation in the neighbourhood of the minimum. Observe the behaviour of the error on the final plot.

4. Choose the momentum method with the value of the momentum greater than zero. Repeat the experiment with different values of momentum for the same initial point. Try to choose the value of the momentum such that no oscillations in the learning progress occur.

5. Compare the plot of the error with the previous one, obtained for standard learning.

Questions

1. How does the momentum method influence the efficiency of learning?

2. Interpret the momentum method on “mechanistic” grounds.

3. Try to explain why the progress of learning is very poor if the momentum is close to 1.

4. Give some rules for momentum choice.

Example 2 – initial point

Use the program `nn1_2` to learn how the Nguyen–Widrow method for initial point choice improves the overall learning progress.

Quick help

1. Enter the program `nn1_2` and input the set of 4–6 points.
2. Choose the Nguyen–Widrow method and observe the initial point calculated by the program.
3. Repeat the experiment for different sets of points to be approximated.

Questions

1. How far is the location of the initial point from the global minimum?
2. Does the location depend on the shape of the error surface?

Example 3 – adaptive learning rate

Adaptation of the learning rate is one of the basic techniques applied in the learning procedure. Use the program `nn1_2` to learn how it works.

Quick help

1. Enter the program `nn1_2` and input the set of points to be approximated in such a way that the final error surface is corrugated.
2. Choose standard learning with default parameters. Find the initial point such that the learning is "unstable", i.e. the trajectory of the error goes outside the error contour plot.
3. Repeat the experiment with an adaptive learning rate. Use the same initial point. Try to find the parameters of the adaptation such that the learning progress is satisfactory. This can be done without changing the number of epochs and error-goal.
4. Verify the method using different values of learning rate increment `Lr_inc`, learning rate decrement `Lr_dec` and error ratio `Err_ratio`. Observe and compare plots of error trajectory and network learning rate.

Questions

1. Give an interpretation of learning rate adaptation as gradient correction.
2. How does the learning rate depend on the shape of the error surface?
3. Evaluate the efficiency of the learning rate adaptation.

4. How do the parameters of the adaptation influence the error minimization? Give some rules for parameter choice.
5. What is the dependence between the number of epochs needed and the adaptation parameters?

Training of the network using the NN Toolbox

The most important part of the program nn1_1 an nn1_2 is network training:

```
W = W0;
B = B0;

A = logsig(W*P,B);
E = T-A;
SSE = sumsqr(E);

for epoch=1:max_epoch

  if SSE < err_goal, epoch=epoch-1; break, end

  D = deltalog(A,E);
  [dW,dB] = learnbp(P,D,lr);
  W = W + dW;
  B = B + dB;

  A = logsig(W*P,B);
  E = T-A;
  SSE = sumsqr(E);

end
```

The first two lines define initial values W0 and B0 of weight W and bias B. The next two calculate output value(s) of the neuron(s) A and initial error E between A and target T. The loop between for and end is repeated max_epoch times or is broken if the current squared error SSE is less than assumed. Within the loop the derivatives D of error are calculated and the back-propagation formula (4.41) is applied. Then weight and bias are updated and the new output and error are calculated.

The same scheme will be used in the examples of Section 4.8.3.

4.8.3 Off-line neural model identification

Off-line identification will be performed by presentation of a predefined number of previously obtained patterns. The patterns are made up of a series of input and corresponding output signals. The approach is much the same as in standard identification (e.g. least squares), the only difference being that presentations of the data are repeated. One presentation is called an epoch. The network consists of one hidden layer and has one output neuron related to model output $y(i)$. The number of input neurons follows from the assumed structure of the plant:

- NY neurons related to delayed output of the plant $y(i-1), y(i-2), \ldots, y(i-NY)$
- $NU+k$ neurons related to delayed input of the plant $u(i–k–1), u(i–k–2), \ldots, u(i–k-NU)$.

The number of hidden neurons nun is chosen by the user. The transfer function of the neuron can be sigmoidal (non-linear network) or linear (linear network). Note that output neuron always have linear transfer function.

Network structure

It is difficult to discuss network structure choice if the plant is non-linear because there is no general method of non-linear model description. In the following exercises, because of the simplicity of interpretation, a linear plant will be tested.

Example 1 – non-linear network structure versus linear plant structure

Use the program `nn2_1` to check the influence of the network structure on the quality of the identification. Let the plant have first-order inertia.

Quick help

Generate data (patterns) for the purpose of identification and validation of neural models:

1. Enter the m-file `nn_in1` at the MATLAB prompt. A window with a SIMULINK diagram appears.
2. Open SIMULINK scope blocks in order to observe the input and output signals.
3. Perform the simulation. At this stage it is not suggested that default values of simulation and input signal be changed. (Note, however, that the input has stochastic character.) In this way input–output data (UY) are generated.
4. Quit the window with `nn_in1`.

Generate data (patterns) validation of the neural model:

1. Enter the m-file `nn_in1s` at the MATLAB prompt. A window with a SIMULINK diagram appears.
2. Open scope blocks in order to observe the input and output signals.
3. Perform the simulation. Note that the step response of the plant is obtained. This is stored in the MATLAB workspace as the response `RES`.

4. Quit the window with `nn_inls`.

Perform the identification and validation experiment.

1. Enter the program `nn2_1`.
2. Choose the network structure.
3. After the plant response presentation, press any key in the MATLAB command window in order to advance the experiment; choose the initial network values. If the program is running for the first time, Nguyen–Widrow initialization is needed. It is possible, however, to store (save) the final network parameters in the MATLAB workspace and then repeat the program with these initial values.
4. The first approximation is shown. To advance the program, press any key in the MATLAB command window.
5. Choose the parameters of the network. It is preferred at this stage to choose the default parameters.
6. According to the value of `disp_freq`, the progress of learning can be observed in the graph window where step responses of the plant and neural model are shown together, and also in the MATLAB command window where current mean error related to the last epoch is displayed. Note that the error is based on the patterns used in identification, while validation can be done by observation of the step responses.
7. After the program reaches `max_epoch`, we can observe the changes of the network learning rate during the experiment by pressing any key in the MATLAB command window. Then the network error is displayed in the graph window. Finally, output vector errors (bar chart) are shown.
8. Repeat the program with different structures of the network. Observe the influence of the structure choice on the quality of identification.
9. We can repeat the program to identify second-order systems. m-file `nn_osc` should be entered; data for identification are obtained. m-file `nn_oscs` provides the step response of the second-order system. The whole experiment can be then repeated using program `nn2_1`.

Questions

1. How does the structure of the network influence the quality of the identification? How do the parameters NY, NU and k influence the quality of identification?
2. What is the convergence of the identification evaluated by means of step response comparison?
3. How does the structure of the network depend on the structure of the plant?
4. Evaluate the minimal number of neurons (in input and hidden layers) which provide satisfactory accuracy of identification. Evaluate the robustness of the network to structure choice.

Exercise 2 – linear network structure versus linear plant

A linear network has linear transfer functions for all neurons involved. Use the program `nn2_2` to check the efficiency of the neural identification of a linear plant.

Quick help

1. Invoke m-files creating data for identification and validation of neural model. Use a first- or second-order system as in the previous example.
2. Enter the program `nn2_2` at the MATLAB prompt.
3. Run the program in the same way as in the previous example. Note the speed of convergence and the accuracy of the identification process.
4. Repeat the program with different choice of network structure.

Questions

1. What is the difference in convergence behaviour between using non-linear and linear networks for identification of the neural model of the linear plant?
2. Is the linear network more accurate?
3. Interpret the results of the experiment: why are there such large differences between the linear and non-linear networks?
4. How does the structure of the network depend on the plant structure? Give the rules for minimal structure choice.
5. Evaluate the robustness of the network to structure change. Which part of the network structure has the greatest influence on identification efficiency: NY, NU, k, nun?

Excitation choice

In the previous experiments the excitation of the plant was a default. Use again programs `nn2_1` and `nn2_2` and the proper m-files producing data for identification and validation, but this time choose a different type of excitation.

Exercise 1 – number of measurements

Quick help

1. Use a stochastic input as in the previous experiment. Do not change the default values of the amplitude (peak).
2. Repeat the program `nn2_1` with different values of the simulation parameter `Stop Time` in m-file `nn_in1` or `nn_osc`. The parameter `Stop Time` defines the number of measurements in the data set used for identification.

3. Repeat in the same way the program `nn2_2` to check the influence of data length on the identification accuracy of a linear network.

4. Increase the maximal number of epochs to improve the results of the identification if the number of measurements is small.

Questions

1. How does the number of measurements influence the accuracy and speed of identification convergence in the linear and non-linear cases?

2. Which property of non-linear network suffers most from the length of the identification data being too short?

3. How does the maximal number of epochs influence the accuracy of identification if the number of measurements is small?

Exercise 2 – amplitude of excitation

Use the programs `nn2_1` and `nn2_2` to check how the amplitude of stochastic excitation influences the quality of identification.

Quick help

1. Use appropriate m-files to generate identification and validation data. Do not change the default values of the data length.

2. Repeat the program `nn2_1` and `nn2_2` with different values of input signal amplitude (peak). Use an input signal in the range (0.1 – 8).

3. Choose the best network structure for the related plant.

4. Do not change the frequency of the input signal.

5. Repeat the experiment for linear and non-linear networks using the same choice of the amplitude.

6. Increase the maximal number of epochs to improve the accuracy of identification if the amplitude is extremely small.

Questions

1. Compare the influence of excitation amplitude on linear and non-linear neural models.

2. Give the rule for the choice of input signal amplitude in both linear and non-linear cases.

3. Interpret the identification results in the non-linear case for excitations of large amplitude.

Exercise 3 – poor excitation

Stochastic excitation is the best for identification. In practice, however, it is often difficult to input such a signal. Use the same programs as in the previous experiments, to check how poor excitation causes the accuracy of identification to deteriorate.

Quick help

1. Use deterministic signals:
   - sine
   - square or step
   - saw-tooth
2. Choose low frequencies (0.1 – 0.5).
3. Because the statical gain of the system considered is 1, choose amplitude 1 and do not change it.
4. Choose the best network structure for the related plant.
5. Repeat the experiment for linear and non-linear networks using the same amplitude.

Questions

1. Compare the influence of poor excitation on linear and non-linear neural models.
2. Why do the greatest output vector errors arise in the first part of the identification data?
3. Which input signal is worst and why?

Learning parameters

In the previous experiments an adaptive learning rate method was incorporated. Use programs `nn2_1` and `nn2_2` to check the influence of the learning parameters on identification quality.

Exercise 1 – initial learning rate

For the non-linear network, evaluate the influence of initial learning rate on identification efficiency.

Quick help

1. Use the example of a second-order system (m-files `nn_osc, nn_oscs`).
2. Perform the first experiment for default values of the parameters. Only the maximum number of epochs can be reduced in order to make whole experiment faster.
3. Repeat the program `nn2_1` using different values of the initial learning rate.
4. Observe the speed of convergence as shown by the quality of the step-response approximation and network error.
5. Pay special attention to the behaviour of the learning rate during identification.

Questions

1. Has the change of initial learning rate any influence on the speed of convergence?
2. Give rules for determining the initial value of the learning rate.
3. How can the results be interpreted?

Exercise 2 – adaptation parameters

The increment of learning rate `lr_inc`, decrement of learning rate `lr_dec` and error ratio `err_ratio` are the parameters of adaptation. Use program `nn2_1` to check their influence on the convergence of learning.

Quick help

1. Use an example of a second-order system (m-files `nn_osc, nn_oscs`).
2. Perform the first experiment for default values of the parameters. Only the maximum number of epochs can be reduced to make the whole experiment faster.
3. Repeat program `nn2_1` using different values of the adaptation parameters. Begin with a large value of initial learning rate and stop the adaptation by entering increment and decrement equal 1.
4. Make `lr_inc` greater than its default value and `lr_dec` close to but less than 1. This provides a tendency to increment the learning rate much more rapidly than decreasing it.
5. Repeat the experiment above with the opposite tendency: small `lr_dec` and `lr_inc` close to but greater than 1.
6. Perform an additional experiment with a different `err_ratio`.
7. Observe the speed of convergence as shown by the quality of the step-response approximation and network error.
8. Pay special attention to the behaviour of the learning rate during identification.

Questions

1. Which of the three adaptation parameters has the greatest influence on the identification results?

2. Give an interpretation of the influence of the adaptation parameters on identification quality.

3. Formulate rules for choosing the adaptation parameters.

Non-linear plant identification

Neural models offer the possibility of compact description of any non-linear plants. In the following, exercises are provided showing problems of neural model identification of a non-linear plant. The plant is a well-known simplified discrete-time description of a stirred-tank reactor:

$$y(i) = \frac{y(i-1)}{1+y^2(i-1)} + u^3(i-1)$$

Note both dynamical non-linearities and the non-linear statical characteristic of the plant.

Exercise 1 – pattern choice

One of the main differences between linear and non-linear plants is that the characteristics of the linear plant do not depend on the input signals. Thus, validation of a model can be done, for instance, by comparison of step responses. For non-linear plants, a specific input signal has to be chosen in order to generate an output pattern. The pattern will be then used for model validation. Use m-file `nn_nl1s` to generate validation data.

Quick help

1. Enter m-file `nn_nl1s` at the MATLAB prompt.

2. Perform the experiment several times using different excitation signals. Begin with steps using different amplitudes. Observe the plant gain dependence on the amplitude of the input signal. Instead of a step function we can use a square wave produced by a signal generator. A sine or saw-tooth wave can be also introduced.

3. Deterministic signals with different amplitudes can be inputted by repeating the sequence block. Evaluate the plant excitation with these signals.

Questions

1. Describe the dependence between the amplitude of excitation and the statical/dynamical properties of the plant.

2. Which of the signals under consideration is the best for validating the model?

3. Give a rule for choosing input signal to produce a good pattern for validation.

Exercise 2 – network structure

Use program nn2_3 to check the influence of network structure choice on identification quality.

Quick help

1. Enter m-file nn_nl1 and generate data for identification.

2. Enter m-file nn_nl1s and generate data for validation.

3. Enter program nn2_3. The program works like nn2_1 and nn2_2. At the beginning, the structure of the network should be chosen, then the validation pattern appears in the graph window, and network initialization values should be introduced. At the very beginning Nguyen–Widrow initialization should be chosen. When the program has ended, the final values of the network parameters can be stored and recalled in the next run of the program.

4. After plotting, the initial response of the network parameters is displayed. It is important that you do not change the default parameters at this stage. Note that the response of the model is calculated for the same excitation signal as in m-file nn_nl1s which produced the validation pattern.

5. During the learning phase, pay special attention to convergence.

6. Repeat the experiment for different network structures.

Questions

1. In the linear case, the optimal choice of network structure relates to the structure of the plant, e.g. the number of delayed samples of plant output NY is equal to the plant dynamical order. Can we apply the same rule in the non-linear case?

2. Which parameter of the network has the greatest influence on the quality of identification?

3. Formulate a rule for choice of network structure for non-linear plant modeling.

4. What is the most significant difference between the network errors in the linear and non-linear cases?

Exercise 3 – excitation choice

Exercise 1 in this section shows that the choice of input signal plays a crucial role in validation. A similar or even more important difficulty concerns the choice of input signal to generate identification data. Use program nn2_3 to check the influence of the input signal choice on identification quality.

Quick help

1. Enter m-file `nn_nl1s` to generate validation data.
2. Enter m-file `nn_nl1`. Choose a sine wave instead of noise as the input signal.
3. Run program `nn2_3`. Observe the quality of identification by comparing plant and model responses.
4. Repeat the program for a square wave. Check for different amplitudes of the square wave.

Questions

1. Why does poor excitation cause such deterioration of the identification results?
2. Give an interpretation of the identification bias due to the amplitude of the excitation signal being too small.
3. Does changing the network parameters to provide adaptation of the learning rate improve the learning process if the excitation signal is poor?
4. What is the connection between the structure of the network and the result of identification if the excitation signal is poor?

Exercise 4 – linear network

Using a linear network for identification of a non-linear plant is a kind of linearization. Use the program `nn2_4` to check how a linear network can describe a non-linear plant.

Quick help

1. Run m-files `nn_nl1` and `nn_nl1s` to generate identification and validation data.
2. Enter program `nn2_4`. Repeat the program for different structures and different sets of parameters entered.
3. Observe the ability of the network to approximate the statical/dynamical properties of the plant.

Questions

1. What is the crucial difference between non-linear and linear networks with respect to their ability to describe non-linear plants?
2. Explain why a linear network is not able to approximate the plant.
3. Answer question 1 with respect to amplitude of excitation.
4. Explain the difference between the learning rate adaptation and network error for linear and non-linear networks.

4.8.4 Recursive identification

In adaptive control, the model has to be identified on-line in a recursive way. The following section provides exercises concerning on-line identification of neural models. The structure of the neural model refers to the single-layer discrete-time feedback network described in Section 4.3. It is assumed that the discrete dynamical relation between input u and output y is

$$\begin{aligned} y(i) \quad = \quad & f(y(i-1), y(i-2), \ldots, y(i-nA), \\ & u(i-1-k), u(i-2-k), \ldots, u(i-nB-k)) \end{aligned} \tag{4.85}$$

where i is discrete time and f is an unknown, generally non-linear function. The number of input neurons is then $nA + nB$. k refers to additional discrete-time delay. One hidden layer consists of nun neurons. One output neuron refers to $y(i)$.

Basic properties of neural model recursive identification

Exercise 1 – constant learning rate

Use m-file `nn3_1` to see recursive identification of a linear model.

Quick help

1. Enter m-file `nn3_1` at the MATLAB prompt and open all scope blocks.
2. Open the block `NEMOL` and enter the parameters of the network. Note that the plant is continuous, but the network samples input and output signals, so a discrete-time model is identified.
3. Choose a square wave as the input signal and perform a few experiments with different amplitudes and frequencies of the input signal. Change the excitation parameters during simulation.
4. Repeat the above experiments with a sine wave as the input.

Questions

1. Compare the convergence of the neural-identification with the RLS method.
2. Why is the response of the model asymmetric?
3. Explain why the plant is non-stationary and how the non-stationarity influences the identification error?
4. How can you make the plant stationary?

Exercise 2 – abilities to track changes of the plant

Use m-file `nn3_2` and perform a few experiments with different excitation signals and/or plants. During the simulation, after convergence, change some parameters of the plant.

Quick help

1. Enter m-file `nn3_2` at the MATLAB prompt. Notice that the excitation signal is sampled and held constant between sampling points. Remember to keep the same sampling time for unit delay as in `NEMOL` sampling.

2. Use a learning rate of 0.05.

3. After about 150 sampling periods, change some parameters of the plant, observing the convergence of identification.

4. Repeat the experiments with different learning rates.

Questions

1. Compare the convergence of the neural-identification process with that of parametric models and the RLS method.

2. How does the learning rate influence tracking plant changes?

3. To which parameter change of the linear plant is the neural model most sensitive?

Exercise 3 – structure of the network

Use m-file `nn3_2` to learn how the choice of the network structure influences the quality of recursive identification. Recall the results of the same exercise concerning non-recursive identification.

Quick help

1. Enter m-file `nn3_2` at the MATLAB prompt.

2. Choose the simulated plant as first-order inertia model.

3. Perform a few experiments choosing different structures of the network. Observe which is the best.

4. Repeat the experiment for the simulated plant begin the second-order inertia model.

Questions

1. Which parameter of the network influences the identification accuracy most?

2. How does the best choice of the network structure depend on the plant structure?

3. Formulate a rule for choosing the network structure.

Exercise 4 – non-stationary plant

It is shown in the exercise that improper settings of sampling time for the model and input signal sample-and-hold result in significant deterioration of on-line identification.

Quick help

1. Enter m-file nn3_2 at the MATLAB prompt.

2. Choose first-order inertia as the plant. Keep the network parameters on their default values. For instance, keep the sampling interval of the input signal hold equal to 1 with offset 0.

3. Start the simulation. After convergence is observed, increase the sampling interval to 1.5.

4. Perform some more experiments with different sampling time intervals.

Questions

1. Why does the difference between sampling interval of the hold and the identifier cause such large deterioration?

2. Why is the observed effect especially important in non-recursive identification?

3. Is correct identification possible if we cannot hold the input signal constant between sampling points?

Non-linear on-line identification

The same plant as in the non-recursive approach is considered:

$$y(i) = \frac{y(i-1)}{1+y^2(i-1)} + u^3(i-1).$$

Exercise 1 – choice of learning rate

Use m-file nn3_3 to see how the constant learning rate influences on-line non-linear identification.

Quick help

1. Enter m-file nn3_3 at the MATLAB prompt. Open all scope blocks.

2. Use a sine wave as the excitation signal. The structure of the network is chosen to be: $NY = 5$, $NU = 5$, $k = 0$, $nun = 5$.

3. Use a constant learning rate. Thus `lr_inc`, `lr_dec` and `error_ratio` are kept equal to 1. Hence the adaptation window `lr` does not play a role.

4. Repeat the experiments for `lr` = 0.2, 0.1, 0.05, 0.01. Use 1000 steps in each. Observe the convergence and final approximation.

Questions

1. Which of the learning rates proposed is the best?
2. What effects can be observed if the learning rate is too high or low?
3. Explain learning rate adaptation using the same "epoch" approach as in non-recursive identification.

Exercise 2 – structure choice

Use m-file `nn3_3` to learn how the choice of the network structure influences the quality of recursive identification.

Quick help

1. Enter m-file `nn3_3` at the MATLAB prompt. Open all scope blocks.
2. Enter a constant learning rate of 0.05.
3. Perform the experiment, changing the structure of the network. Try 1000 steps.
4. Observe the convergence and final approximation.

Questions

1. Which of the structure parameters influences the quality of recursive identification most?
2. Compare the conclusion just obtained with the one for non-recursive linear and non-linear identification of the neural model.
3. Give a rule for structure choice.

Exercise 3 – adaptive learning rate

Use m-file `nn3_3` to learn how adaptation of the learning rate improves the results of recursive identification.

Quick help

1. Enter m-file `nn3_3` at the MATLAB prompt. Open all scope blocks.
2. Use a sine wave as the excitation signal. Choose the network structure: $NY = 5$, $NU = 5$, $k = 0$, $nun = 5$.
3. Choose as the parameters of the learning rate adaptation: initial value 0.1, `lr_inc` = 1.05, `lr_dec` = 0.7, `err_ratio` = 1.04. Use one or two periods of the excitation signal as the window for adaptation.
4. Observe the quality of adaptation versus learning rate changes.
5. Repeat the experiment using different values of the adaptation parameters.

Questions

1. Compare the results with the previous one obtained for a constant learning rate. Is any improvement observed?
2. How do the parameters of the learning rate adaptation influence the quality of recursive adaptation?
3. Why should `lr_inc` be close to 1 and `lr_dec` significantly less than 1?
4. Give an interpretation of the influence of `error_ratio` and `window for adaptation` on the change in learning rate.

Exercise 4 – change of the excitation

Up to now a periodic excitation has been used. Use m-file `nn3_3` to learn how the change of excitation type influences the quality of the identification.

Quick help

1. Enter `nn3_3` at the MATLAB prompt and open all scope blocks.
2. Use a sine wave as the excitation signal. Choose as the structure of the network: $NY = 5$, $NU = 5$, $k = 0$, $nun = 5$.
3. Choose as the parameters of the learning rate adaptation: initial value 0.1, `lr_inc` = 1.05, `lr_dec` = 0.7, `err_ratio` = 1.04. Use one or two periods of the excitation signal as the window for adaptation.
4. Start the simulation. After convergence, change the excitation to a saw-tooth or a square wave. Observe the deterioration of identification. Return to the previous excitation and observe the quality of approximation.
5. Repeat the experiments, changing excitation frequency.

Questions

1. How does the change of excitation type influence identification?
2. What role does adaptation window length play if excitation type/frequency changes?
3. Give an idea how to reduce the sensitivity of the network to excitation change?

Software-related notes

On-line identification of neural models requires certain calculations to be performed. The calculations in all examples presented is done by S-functions. Feedback networks are dynamical systems, so S-functions are a convenient way to describe the network as well as perform learning.

Consider S-function `nemo1.m` which describes and performs learning of a one-hidden layer discrete feedback network. The states of the network consist of delayed measurements of the input and output of the plant (feedback within the network), values of the weights and biases of the network, and the outputs of the hidden layer and the network error (the learning process is also dynamical). `nemo1.m` has the following states:

- plant output $y(i-1), \ldots, y(i-nA)$; addresses within state vector: $1...nA$,
- plant input $u(i-1), \ldots, y(i-nB-k)$; addresses within state vector: $nA+1...n$, where $n = nA + nB + k + 1$,
- weights of network hidden layer `W1`; addresses within state vector: $n{+}1...n(nun{+}1)$, where nun is number of neurons in the hidden layer.
- biases of network hidden layer `B1`; addresses within state vector: $n(nun+1) + 1...(n+1)nun + n$,
- weights of network output neuron `W2`; addresses within state vector: $(n+1)nun + n + 1...(n+2)nun + n$,
- bias of network output neuron `B2`; address within state vector: $(n+2)nun + n + 1$,
- outputs of network hidden layer `A1`; addresses within states vector: $(n+2)nun + n + 2...(n+3)nun + n + 1$,
- output of network output neuron `A2`; address within state vector: $(n{+}3)nun{+}n{+}2$,
- network output error `E`; address within state vector: $(n+3)nun + n + 3$.

The parameters of the network and learning procedure (here with a constant learning rate) are entered by the user and passed into the S-function:

1. `nA` number of delay steps of output signal in the model,
2. `nB` number of delay steps of input signal in the model,
3. `k` number of additional delay steps of input signal in the model,
4. `nun` number of neurons in the hidden layer,
5. `lr` learning rate of back-propagation algorithm,
6. `offset` and `ts` offset and sampling time.

The other parameters of the S-function are standard:

- `t` time,
- `x` state vector (defined above),
- `u` input vector: `u(1)` means plant output, `u(2)` means plant input,
- `flag` controls how the S-function is called.

`n` is defined in the preamble of the S-function parameters. At the beginning the S-function is called with `flag` = 0 in order to define its structure and the initial values of the states. Thus return vector `sys` is defined as follows:

- first 0 is no continuous states,
- $n + (n + 3) * nun + 3$ means the number of discrete states,
- next 1 means that the number of outputs of the S-function is 1,
- 2 is the number of inputs,
- the two last zeros mean that there are no discontinuous roots and no direct feedthrough.

If there exist global variables `W10`, `B10`, `W20`, and `B20` in the MATLAB workspace, they are entered as initial values of respective weights and biases, otherwise the weights and biases are initialized as random. The outputs of the neurons as well as the network error are also initialised.

At every sample the S-function is called with `flag` = 2. Old values of the states (vector `x`) are updated (in vector `sys`). The learning procedure is much the same as presented in Section 4.8.2. Note that the network is linear, thus the transfer function is `purelin`. Note also that state updating refers not only to the network parameters but also to the plant input and output signal (measurements).

For `flag` = 3, the output of the network is returned. The input neurons are defined by matrix `P`. The value of the linear neuron output is calculated using the updated network parameters.

For `flag` = 4, the next sample hit is calculated and this completes the calculation within one sample period.

4.8.5 Neural networks for control

Perfect cancellation

Two examples of plants are considered. The previously shown reactor model (see for instance Section 2) and the Chen model:

$$y(i+1) = p_1 \sin(p_2 y(i)) + p_3 u(i)$$

Default values of the parameters are: $p_1 = 0.8$, $p_2 = 2$, $p_3 = 1.2$.

Exercise 1 – dynamics of the Chen example

Use m-file `nn4_1` to learn the dynamical behaviour of the Chen model.

Quick help

1. Enter m-file `nn4_1` at the MATLAB prompt. Open the scope block.
2. Enter default parameters of the plant.
3. Perform a few experiments observing the system response for different excitation signals (sine, square, saw-tooth). Change the amplitude of the excitation during the experiments.

Questions

1. How does the amplitude of the excitation influence the system response?
2. Interpret the oscillatory behaviour of the system.
3. Evaluate the difficulty of controlling the system control.

Exercise 2 – adaptive generalized predictive control

Use m-file `nn4_2` to learn how adaptive GPC controls the system.

Quick help

1. Enter m-file `nn4_2` at the MATLAB prompt. Open the scope blocks.
2. Enter default parameters of the plant.
3. Enter parameters of the controller as follows: $N_1 = 1$, $N_2 = 3$, $N_u = 1$, $r = 0$, $\alpha = 0$.
   N_1 and N_2 are the minimal and maximal horizons, respectively, for output prediction, N_u is the horizon for control signal prediction. Parameter r is the control signal weight in the criterion function and α is the parameter of the first-order filter which defines the reference trajectory ($0 \leq \alpha \leq 1$).
4. Enter the model parameters as follows: $k = 1$, $B = 1$, $A = [1, 0]$.
   k is model discrete-time delay. Vectors A and B contain initial values of the coefficients of the polynomials A and B of the model transfer function.
5. Enter the identification parameters: initial values of diagonal of the covariance matrix $P_{init} = 1000$, forgetting factor $\lambda = 1$.
6. Perform the simulation, stopping if instability occurs.
7. Make the model order greater, e.g. $B = [1, 0, 0, 0]$, $A = [1, 0, 0, 0]$, and repeat the simulation.
8. Make the control weighting factor greater than 0, e.g. $r = 1$, and repeat the simulation.
9. Repeat the simulation with forgetting factor less than 1: $\lambda = 0.95$.
10. Repeat the simulation with the sine reference signal.

Questions

1. Why is the closed-loop system unstable in the first simulation experiment?
2. Why does increasing the model order and control weighting make the closed-loop system stable.
3. Does reducing of the forgetting factor improve the closed-loop system performance? If yes, explain the reason (the plant parameters do not change!).
4. Why is the closed-loop system perform unsatisfactory with a sine reference signal despite its good performance with a square-wave reference?

Exercise 3 – PI control

Use m-file `nn4_3` to learn how a PI controller performs with the Chen example.

Quick help

1. Enter m-file `nn4_3` at the MATLAB prompt. Open the scope blocks.
2. Enter default plant parameters.
3. Perform a few experiments trying to find controller parameters making the performance of the closed-loop system satisfactory.

Questions

1. Compare the quality of PI and GPC control.
2. Which set of parameters is best?
3. Is the PI controller satisfactory?
4. Why does the plant cause such difficulty?

Exercise 4 – perfect cancellation neural controller

Use m-file `nn4_4` to learn how a perfect-cancellation neural controller performs on the Chen example. The neural controller uses two networks "f" and "g" to describe two parts of the plant. Each network contains two hidden layers of neurons.

Quick help

1. Enter m-file `nn4_4` at the MATLAB prompt. Open the scope blocks.
2. Enter default plant parameters.
3. Enter the structure of the networks: in both cases use 10 neurons in the first and the second layer.
4. Use initial learning rate 0.1 and 0.04 for the "f" and "g" networks respectively.

5. Use default values for the sampling parameters.

6. Perform the simulation with the square-wave reference signal. Observe convergence. Once the performance is satisfactory, i.e. we can judge that the networks have already learned, change the reference signal to sine. Again once the performance is satisfactory, change the parameter(s) of the plant and check if the closed-loop system follows the non-stationarity.

Questions

1. Compare the quality of GPC and PI control.

2. Compare the convergence of the adaptation in the GPC and neural cases.

3. Evaluate the convergence of learning, taking into account experience with off-line neural model identification.

4. Compare the numbers of steps needed for learning with a square-wave and a sine-wave reference signals. Explain the differences.

5. How does the closed-loop system follow the plant changes?

Exercise 5 – network structure

Use m-file `nn4_5` to learn how different network structures influence the quality of control.

Quick help

1. Enter m-file `nn4_5` at the MATLAB prompt. Open the scope blocks.

2. Enter default plant parameters.

3. Enter the network structures: in both cases use less than 10 neurons in all layers.

4. Perform a few experiment using a square-wave reference. Stop the simulation if the results are not satisfactory.

Questions

1. Evaluate the influence of the network structure on the control quality.

2. Is any improvement in the choice of network structure possible?

3. Which structure is the best for a square-wave reference?

Exercise 6 – neural-control of the reactor model

Use m-file `nn4_6` to learn how perfect cancellation performs with the reactor model.

Quick help

1. Enter m-file nn4_6 at the MATLAB prompt. Open the scope blocks.
2. Enter default plant parameters.
3. Enter the network structure: in both cases use 10 neurons in all layers.
4. Perform the simulation, stopping after about 150 steps.

Questions

1. Why is the response of the system unstable?
2. Is it possible to control any plant with the perfect cancellation technique?

Neural predictive control

Exercise 1 – neural control of the reactor model

A non-adaptive controller is used to control the reactor model. The predictive controller is of the following type

$$u(i) = \boldsymbol{q}^T(\boldsymbol{w} - \boldsymbol{y}^0)$$

where $\boldsymbol{q}$ is the gain vector containing weights with which future (predicted) behaviour of the system is taken into account while the control signal is calculated, $\boldsymbol{w}$ is the vector of future reference signal values and $\boldsymbol{y}^0$ is the vector of plant response predictions. The predictions were calculated using a neural model obtained off-line (see Section 2).

Use m-file nn4_7 to see how it works.

Quick help

1. Enter m-file nn4_7 at the MATLAB prompt. Open the scope blocks.
2. Use default values for the controller parameters.
3. Perform the simulation, changing the reference signal from sine wave to square wave and/or saw-tooth.
4. Repeat the simulation with different weights in vector $\boldsymbol{q}$.

Questions

1. Is the closed-loop system stable?
2. Is it possible to choose the weights $\boldsymbol{q}$ in such a way as to obtain perfect cancellation?
3. Give an idea how to choose weights $\boldsymbol{q}$.

4.9 REFERENCES

[1] F. Albertini and E. D. Sontag. For neural networks, function determines form. Neural Networks. (to appear). See also *Proc. IEEE Conf. Decision and Control*, Tucson, December 1992, IEEE Publications, 1992, 26–31.

[2] F. Albertini and E. D. Sontag and V. Maillot. Uniqueness of weights for neural networks. In R. Mammone (ed.), *Artificial Neural Networks with Application in Speech and Vision*, Chapman and Hall, London (to appear).

[3] S. Z. Amari. Mathematical foundations of neurocomputing. *IEE Proc.*, **78**(9), 1443–1463, 1990.

[4] J. A. Anderson and E. Rosenfeld. *Neurocomputing: Foundation of Research.* MIT Press, Cambridge, MA, 1988.

[5] F. C. Chen. Back-propagation neural networks for nonlinear self-tuning adaptive control. *IEEE Control Syst. Mag.*, **10**, 44–48, 1990.

[6] S. Chen and S. A. Billings and C. F. Cowan and P. M. Grant. Practical identification of NARMAX models using radial basis functions. *Int. J. Control*, **52**, 1327–1350, 1990.

[7] C. E. Garcia and M. Morari. Internal model control - I. A unifying review and some new results. *Ind. Eng. Chem. Process Des. Dev.*, **21**, 308–323, 1982.

[8] S. Grossberg. Adaptive pattern classification and universal recording: I. Parallel development and cooling of neural feature detectors. *Biol. Cybernet.*, **23**, 121–134, 1976.

[9] D. O. Hebb. *The Organization of Behaviour.* Wiley, New York, 1949.

[10] R. Hecht-Nielsen. Neurocomputing applications. In R. Eckermiller and Ch.v.d. Malsburg (eds), *Neural Computers*, pp. 445–453. Springer-Verlag, Berlin, 1988.

[11] J. J. Hopfield. Neural Networks and physical systems with emergent collective computational abilities. *Proc. Nat. Acad. Sci. USA*, **79**, 2554–2558, 1982.

[12] K. Hornik and M. Stinchcombe and H. White. Multilayer feed-forward networks are universal aproximators. Discussion paper. Department of Economics, University of California, San Diego, La Jolla, CA, 1988.

[13] K. J. Hunt and D. Sbarbaro. Neural networks for nonlinear model control. *Proc. IEE D*, **138**, 431–438, 1991.

[14] K. J. Hunt and D. Sbarbaro and R. Zbikowski and P. J. Gawthrop. Neural networks for control systems. A survey. *Automatica*, **28**, 1083–1112, 1992.

[15] IEEE. Special issue on neural networks. *IEEE Control Syst. Mag.*, **8**, 1988.

[16] IEEE. Special issue on neural networks. *IEEE Control Syst. Mag.*, **9**, 1989.

[17] IEEE. Special issue on neural networks in control systems. *IEEE Control Syst. Mag.*, **10**, 1990.

[18] IEEE. Special issue on neural networks. *IEEE Control Syst. Mag.*, **12**, 1992.

[19] G. Kulawski and M. A. Brdyś. *Fourth International Conference on Control '94.* University of Warwick, UK, 21–24 March 1994.

[20] J. G. Kuschewski and S. Hui and S. H. Zak. Application of feed-forward neural networks to dynamical system identification and control. *IEEE Trans. Control Syst. Technol.*, **1**(1), 37–49, 1993.

[21] T. L. McClelland and D. E. Rummelhart and the PDP Research Group. *Parallel Distributed Processing.* MIT Press, Cambridge, MA, 1986.

[22] W. S. McCulloch and W. Pitts. A logical calculus of the immanent in nervous activity. *Bull. Math. Biographics*, **9**, 127–147, 1943.

[23] W. T. Miller and R. S. Sutton and P. J. Werbos. *Neural Network for Control.* MIT Press, Cambridge, MA, 1990.

[24] M. L. Minsky and S. A. Pappert. *Perceptrons.* MIT Press, Cambridge, MA, 1969.

[25] K. S. Narendra and K. Parathasarathy. Identification and control for dynamic systems using neural networks. *IEEE Trans. Neural Networks*, **1**, 4–27, 1990.

[26] D. H. Nguyen and B. Widrow. Neural networks for self-learning control systems. *IEEE Control Syst. Mag.*, **10**, 18–23, 1990.

[27] D. Psaltis and A. Sideris and A. A. Yamamura. A multilayer neural network controller. *IEEE Control Syst. Mag.*, **8**, 17–21, 1988.

[28] F. Rosenblatt. The perceptron: a probabilistic model for information storage and organization in the brain. *Psychol. Rev.*, **65**, 386–408, 1958.

[29] D. E. Rummelhart and G. E. Hinton and R. J. Williams. Learning internal representations by error propagation. In D. E. Rummelhart and J. L. McClelland (eds), *Parallel Distributed Processing.* MIT Press, Cambridge, MA, 1986.

[30] D. E. Rummelhart and J. L. McClelland. *Parallel Distributed Processing: Explorations in the Microstructures of Cognition*, Vol. 1: *Foundations.* MIT Press, Cambridge, MA, 1986.

[31] R. M. Sanner and J. J. E. Slotine. Gaussian networks for direct adaptive control. *IEEE Trans. Neural Networks*, **3**(6), 837–863, 1992.

[32] H. J. Sira-Ramirez and S. H. Zak. The adaptation of perceptrons with application to inverse dynamic identification of unknown dynamic systems. *IEEE Trans. Syst. Man. Cybern.*, **21**, 634–643, 1991.

[33] J. J. E. Slotine and R. M. Sanner. Neural networks for adaptive control and identification. In H. L. Tretelman and J. C. Willems (eds), *Essays on Control: Perspectives in the Theory and its Applications*, pp. 381–436. Birkhäuser, Boston, MA, 1993.

[34] K. Warwick and G. W. Irwin and K. J. Hunt. Neural networks for control systems. *IEEE Control Engineering*, Series 46, London, 1992.

[35] B. Widrow and M. E. Hoff. Adaptive switching circuits. 1960 IEEE WESCON Convention Record, New York: IRE, pp. 96–104, 1960.

[36] B. Widrow. Generalization and information storage in networks of adaline neurons. In M. C. Jovitz, G. T. Jacobi and G. Goldstein (eds), *Self-Organizing Systems*, pp. 453–461. Washington DC, Spartan Books, 1962.

[37] B. Widrow and M. A. Lehr. Thirty years of adaptive neural networks: Perceptron, madaline, and backpropagation. *Proc. IEEE*, **78**, 1415–1442, 1990.

5

Adaptive control

5.1 INTRODUCTION

5.1.1 Adaptive control schemes

A real-world plant can be usually characterized by time-varying dynamical properties – most of them as a result of plant non-stationarity, non-linearity and random disturbances which affect the plant behaviour in the following way:

- the plant is called non-stationary if its dynamics changes in time, e.g. as a result of ageing effects after being in operation for long time;
- if the plant is non-linear (as *every* real plant is) then the dynamical properties of its linearized model are different in the vicinity of various steady-state points; in normal operating conditions the steady-state point changes;
- stochastic models are used to represent the disturbances acting at the plant output because of the large number and different nature of the factors disturbing the normal plant operation.

It is clear that the control algorithm used in the above circumstances should either be adaptive or should exhibit some robustness properties with respect to poor plant models and changes in the plant dynamics.

Robustness properties can usually be ensured by the feedback structure of the control system. The feedback compensates for the deviation of the plant output signal value from its set point, no matter which factor has caused such deviation: disturbances affecting the plant, improper plant model structure or a change in the plant model parameters. However, the two latter factors usually cannot be dealt with well enough by the control system feedback structure alone. Large differences in plant model structure and large changes in the plant dynamics may cause the natural robustness properties of the control system to be exhausted – thus causing unacceptable degradation of the system performance.

The adaptive control algorithms can also be used in such complex and difficult environment conditions. The main concept of this type of control algorithms is to ensure such automatic change of the controller structure and parameters so that they correspond to the current properties of the plant and its environment – in order to achieve considerable improvement of the control system robustness. The usual methods of changing the controller parameters are [5]:

- programmed changes of the controller parameters, also known as *gain scheduling*;

or

- identification of a plant model.

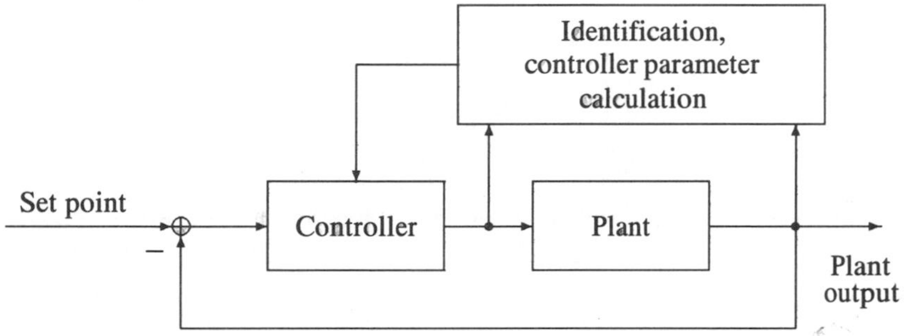

Figure 5.1. Block diagram of an adaptive control system with plant model identification

In this chapter we are going to consider the second type of adaptive control algorithm. The block diagram of such a control system is presented in Fig. 5.1. The controller parameters are calculated as a result of recursive identification of the appropriate plant parametric model, performed on-line.

It is obvious that this control system structure has the potential to capture all changes in the plant parameters – no matter what their origin. We should also mention that the scheme has a rather important drawback: it is extremely complicated with respect to its theoretical analysis. The complexity of the theoretical analysis results mainly from non-stationarity, non-linearity and stochastic disturbances. This is one of the primary reasons that why we are going to use the MATLAB/SIMULINK software tools to investigate such adaptive systems.

We shall now say what kind of model we are going to identify in the adaptive control system. There are two possibilities:

1. A model of the plant to be controlled – the block diagram explaining this kind of adaptive control with plant model identification is presented in Fig. 5.2. On the basis of the plant model and control criterion, we can proceed with control synthesis, i.e. calculate the parameters of the controller. This scheme is called *indirect adaptive control*, because to find the proper values of the controller parameters we have to complete the intermediate plant model identification task.

2. It may be possible to identify the parameters of the *controller* that we are seeking. This scheme is called *direct adaptive control*, because we are going to obtain directly the required controller parameters through their estimation in an appropriately redefined plant model. The block diagram of such an adaptive control system is presented in Fig. 5.3.

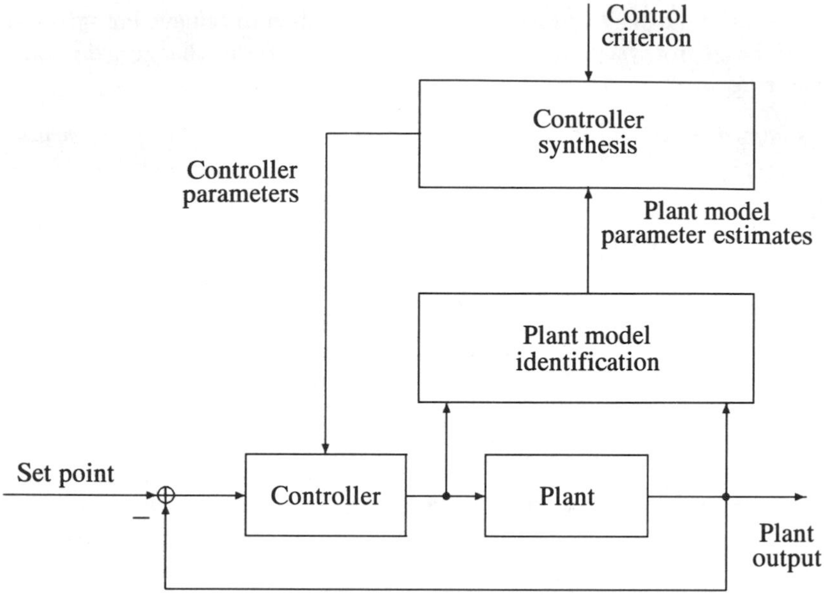

Figure 5.2. Block diagram of an indirect adaptive control system

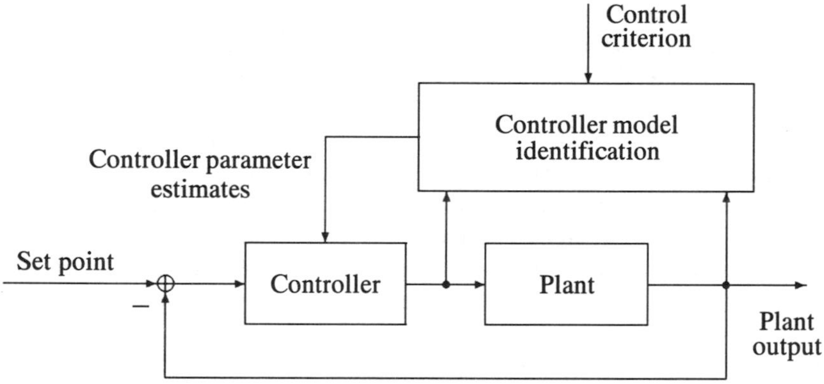

Figure 5.3. Block diagram of a direct adaptive control system

Both methods are known to have several important advantages as well as drawbacks. The main advantages of the *indirect* adaptive control scheme include the general usefulness of the plant model obtained from identification; the model may be used for controller synthesis with several different control algorithms in mind; another useful property of this scheme is the possibility of starting identification while the plant is controlled by virtually any stabilizing controller (for direct adaptive control the controller parameter identification is possible only if the currently tuned controller is working in the feedback channel of the system). On the other hand, identification may be performed by simpler and more robust methods in the *direct* adaptive control system, and as the controller synthesis has to be done only once, we can do it off-line.

In what follows we shall assume that direct adaptive control is used for minimum-variance and pole/zero-placement controllers, while the indirect adaptive control scheme will be chosen for long-range predictive control algorithms.

5.1.2 Plant model

Adaptive control is usually used to cope with an *unknown* or/and *changing* plant to be controlled. Analysis and synthesis of such a control system is possible only under some assumptions concerning the nature of the plant and its dynamics. In this chapter only linear, discrete-time plants disturbed in a deterministic or stochastic manner will be considered. The following plant model will be used:

$$y(i) = z^{-k}\frac{B}{A}u(i) + \frac{C}{A\nabla^{\ell}}e(i) + d(i) + b \tag{5.1}$$

Model (5.1) is one of the most typical in the field of adaptive control and non-standard discrete-time control algorithms in general [4, 5].

The part $z^{-k}\frac{B}{A}u(i)$ in (5.1) will be referred to as the control channel of the plant. If

- $B = B^{+}$ (only stable factors exist in the B polynomial), the plant will be called *minimum phase* (MP),
- $B = B^{+}B^{-}$, the plant will be called *non-minimum phase* (NMP).

In the discrete-time models of the plants, NMP zeros tend to be very common, mainly because:

- NMP zeros to the left of the unit circle in the z-plane are generated as a result of:
  - lack of synchronization of the plant output sampling and changes in the control signal;
  - the presence in the continuous-time model of the plant of a time delay which is not an integer multiple of the sampling interval;
  - choice of too small a sampling interval;
  - choice of too large a discrete-model time delay for model parameters estimation;
  - assumption of too large a number of poles compared with the number of zeros in the continuous-time plant model.
- NMP zeros to the right of the unit circle in the z-plane are generated as a result of the presence of NMP zeros in the continuous-time model of the plant, i.e. the plant itself is NMP (not only its discrete-time model).

The part $\frac{C}{A\nabla^{\ell}}e(i) + d(i) + b$ in (5.1) is called the disturbance channel of the plant, with $\frac{C}{A\nabla^{\ell}}e(i)$ being the stochastic part of the disturbance, $d(i)$ the deterministic part, and b a constant bias of the plant output (deterministic in nature); the three parts represent all disturbances affecting the plant output.

Stochastic disturbances are modelled as the output of a stable, invertible linear filter with the transfer function $\frac{C}{A\nabla^{\ell}}$. The filter is assumed to be excited with white noise $e(i)$

of variance λ^2. Such disturbance is stationary for $\ell = 0$. For $\ell \geq 1$ the disturbance is non-stationary, i.e. it exhibits, for example, a changing mean value or ramp with changing slope, with its ℓth difference being still stationary.

The following two kinds of C polynomials in the disturbance channel should be treated differently:

1. C polynomials passing the strict *positive realness* test, i.e.

$$\mathrm{Re}\left\{\frac{1}{C} - \frac{1}{2}\right\}\Bigg|_{z=e^{j\omega T_s}} > 0 \quad \text{for all } 0 \leq \omega T_s \leq \pi \tag{5.2}$$

2. C polynomials violating the strict *positive realness* condition, i.e.

$$\mathrm{Re}\left\{\frac{1}{C} - \frac{1}{2}\right\}\Bigg|_{z=e^{j\omega T_s}} \leq 0 \quad \text{for some frequencies } 0 \leq \omega T_s \leq \pi \tag{5.3}$$

The *positive realness* condition is often essential while proving convergence of estimation schemes in the adaptive control system, e.g. recursive least squares in direct adaptive control, or extended least squares method in indirect adaptive control.

It is generally agreed that the dynamical model, representing the *deterministic* disturbance as the response of a linear filter excited with the Kronecker delta function, assuming zero initial conditions, is the most suitable one for plant modelling:

$$d(i) = \frac{d}{A_g}\delta(i) \tag{5.4}$$

with

$d(i)$ – deterministic disturbance at the output of the plant;

d – disturbance amplitude;

$A_g = A_g(z^{-1})$ – deterministic disturbance generating polynomial;

$\delta(i) = \begin{cases} 1 & \text{for } i = 0 \\ 0 & \text{for } i \neq 0 \end{cases}$ – Kronecker delta function.

The zeros of the generating polynomial A_g determine the basic properties of the signal:

- zeros on the unit circle in the z-plane imply a non-vanishing disturbance signal;
- zeros inside the unit circle generate a vanishing deterministic disturbance;
- zeros outside the unit circle produce a bursting disturbance signal.

The constant bias b of the plant output is a special kind of deterministic disturbance; we may alternatively generate it using the polynomial $A_g = 1 - z^{-1}$. As the simplest kind of deterministic disturbance it was specially modelled in (5.1), and some special procedures for damping disturbances of this kind are also used in adaptive control system synthesis.

5.2 CONTROL ALGORITHMS SUITABLE FOR ADAPTIVE SYSTEMS

In this section some of the most popular adaptive control algorithms will be presented. The spectrum of adaptive control algorithms described here does not pretend to be complete, rather some interesting adaptive control schemes also used in practice have been chosen.

Generally, most control algorithms presented here could be described by the structure and parameters of the difference equation:

$$R(z^{-1})u(i) + S(z^{-1})y(i) - T(z^{-1})w(i) + h = 0 \tag{5.5}$$

where

$u(i)$ – plant control signal,

$y(i)$ – plant output signal,

$w(i)$ – set point of the plant output signal,

h – constant term.

The coefficients of the $R(z^{-1})$, $S(z^{-1})$ and $T(z^{-1})$ polynomials and the h term are chosen by the user before the simulation experiment and stay constant during the experiment.

5.2.1 Minimum-variance control algorithms

The aim of the minimum-variance control algorithm is the minimization of the following performance index [4, 5, 21]:

$$J = \mathrm{E}\left\{\left[P(z^{-1})\frac{\tilde{B}^-(z^{-1})}{B^-(z^{-1})}y(i+K) - V(z^{-1})w(i)\right]^2 + q\left[Q(z^{-1})u(i)\right]^2\right\} \tag{5.6}$$

where $P(z^{-1})$, $Q(z^{-1})$ and $V(z^{-1})$ are filter polynomials affecting the minimized performance index signal components, q is a weighting coefficient for the control signal value $u(i)$, $\tilde{B}^-(z^{-1})$ is the reciprocal polynomial of the $B^-(z^{-1})$ polynomial, K is the prediction horizon in the performance index.

Minimization of this performance index leads to a control algorithm of the same structure as (5.5). All parameters in the controller equation are identifiable from the prediction model (assuming $K = k$):

$$\begin{aligned}\phi(i) - RB^-u(i-k) - SB^-y(i-k) + TB^-w(i-k) - hB^-(1)\\ = [1-C][\phi(i) - Fe(i)] + Fe(i)\end{aligned} \tag{5.7}$$

where

$$\phi(i) = P\tilde{B}^-y(i) + \frac{q}{b_0^+}QB^-u(i-k) - VB^-w(i-k) \tag{5.8}$$

$$dR = \max(dB + k - 1, dQ + dC)$$

$$dS = \max(dA - 1, dP + dC - k)$$

$$dT = dV + dC$$

The generalized minimum-variance control algorithm presented above ensures that the resulting control system is stable for the *non-minimum phase* plant, even without having to use a non-zero q weighting coefficient and the $Q(z^{-1})$ polynomial, filtering the control signal in the minimized performance index (control signal value weighting is the standard way of stabilizing the minimum-variance control system with NMP plant, but far from optimum).

Choice of minimum-variance control parameters

The following parameters of the minimum-variance control algorithm specification may be adjusted to guarantee the desired properties of the control system:

- the weighting coefficient for the control variable of the plant, which allows "soft" saturation of the control signal value and makes it possible to use this type of control algorithm for some NMP plants;
- the filter polynomial for the plant control signal, which allows the same kind of control signal saturation as described above, thus making it possible to use the minimum-variance control algorithm for some NMP systems; it also enables integral action to be introduced into the control system;
- the filter polynomials for the plant output signal and its set point in the performance index, which allow the introduction of integral action into the control system loop and the possibility of specifying a chosen model of set point followed by the plant output signal (modified tracking properties).

5.2.2 Pole/zero-placement algorithms

As an example of pole/zero-placement control algorithms, some basic results for pole/zero-placement control of a NMP plant will be presented. The control algorithm is synthesized under the assumption that the control aim is to achieve the following transfer function (relating set point changes to plant output changes, cf. e.g. [4, 5, 21]):

$$K_m(z^{-1}) = z^{-k_m} \frac{K_m B^-(z^{-1}) B_m(z^{-1})}{A_m(z^{-1})} \tag{5.9}$$

where k_m is the discrete-time delay in the desired model of set point following, $A_m(z^{-1})$ and $B_m(z^{-1})$ are the polynomials determining the desired poles and zeros of the $K_m(z^{-1})$ transfer function, and the scalar coefficient K_m is

$$K_m = \frac{A_m(1)}{B^-(1)} B_m(1)$$

The effect of random disturbances on the plant output may be diminished by using a special form of observer polynomial $A_o(z^{-1})$, in the pole/zero-placement control synthesis:

$$A_o(z^{-1}) = C(z^{-1}) A'_o(z^{-1})$$

where $A'_o(z^{-1})$ is the basic observer polynomial, used mainly for adjusting the robustness properties of the resulting control system.

The parameters of the synthesized controller equation may be identified from the prediction model (assuming $k_m = k$):

$$\phi(i) - B^- Ru(i-k) - B^- Sy(i-k) + B^- Tw(i-k) - B^-(1)h \\ = [1-C]\left[\phi(i) - R'e(i)\right] + R'e(i) \tag{5.10}$$

where

$$\phi(i) = A'_o A_m y(i) - A'_o B_m K_m B^- w(i-k) \tag{5.11}$$

$$dR' = k - 1 + dB^-$$

$$dR = k - 1 + dB$$

$$dS = \max(dA - 1, dA_m + dA_o - k - dB^-)$$

$$dT = dB_m + dA_o$$

Choice of pole/zero-placement control parameters

The following parameters of the pole/zero-placement control algorithm may be adjusted to achieve the desired properties of the resulting control system:

- the polynomials defining the poles of the control system transfer function, determining the basic properties of the transient signals after set point changes and disturbance changes;
- the polynomials defining the zeros of the control system transfer function, modifying the tracking properties of the plant under control (only the stable part of the original plant $B(z^{-1})$-related dynamics can be modified for NMP plants).

5.2.3 Simple self-tuning control algorithms

A simple self-tuning control algorithm has been proposed by Åström [3]. This is essentially one of the pole-placement control algorithms, derived under the assumption that the plant under control may be described by the following simple model:

$$y(i) = -a_1 y(i-1) - a_2 y(i-2) + b_0 u(i-1) + b_1 u(i-2) + d \tag{5.12}$$

It is also assumed that the poles of the transfer function relating the changes in the set point signal with the changes in the plant output signal should be determined with the following formula:

$$z = e^{-\xi\omega_0 T_p}\left[\cos\left(\omega_0 T_p\sqrt{1-\xi^2}\right) \pm j\sin\left(\omega_0 T_p\sqrt{1-\xi^2}\right)\right]$$

thus corresponding to the poles of the following Laplace transform transfer function:

$$s = -\xi\omega_0 \pm j\omega_0\sqrt{1-\xi^2}$$

where ξ is the plant damping ratio, and ω_0 is the natural oscillation frequency of the plant.

The parameters of this control algorithm may be easily identified from the properly configured prediction model.

This kind of adaptive control algorithm has become quite popular because of its similarity to the well-known, traditional design procedure for PI/PID controllers based on a simplified model of the plant (first or second order) and links to the family of pole-placement control algorithms. The two parameters of the simple self-tuning control algorithm as presented here, the relative damping and natural oscillation frequency, can easily be interpreted in terms of the plant step-response properties.

Choice of the simple self-tuning controller parameters

The following parameters of the simple self-tuning control algorithm may be changed in order to influence the basic set-point-following properties of the resulting control system:

- control system damping ratio – allows determination of the damping ratio affecting the amplitude of the transient signals in the control system, after a set point change or in the presence of disturbances;
- transient signal natural oscillation frequency – allows determination of the frequency of oscillations in the control system transient signals, after a change of the set point value or in the presence of disturbances (the oscillation period should be measured in sampling intervals).

5.2.4 MAC control algorithm

Some basic predictive indirect adaptive control algorithms will be described below, including the multistep model algorithmic control (MAC) presented here and the generalized predictive control algorithm (GPC) described in the next section.

The MAC control algorithm derivation is based on the assumption that the plant under control may be described with a weighting function approximate model, and that the plant is undisturbed, i.e. $e(i) = 0$. The basic control system performance assumption, taken into consideration while deriving the MAC algorithm, is that the response of the plant output signal to a set point change should be the same as the response of the prespecified first-order system of the form:

$$y^Z(i) = z^{-k}\frac{1-\beta}{1-\beta z^{-1}}w(i) \tag{5.13}$$

where $y^Z(i)$ is the desired trajectory of the plant response to the $w(i)$ set point changes, and β determines the speed with which the change is followed.

The control algorithm aims at minimizing the following performance index (corresponding to the above control system performance formulation):

$$J = \sum_{j=0}^{H}\left[\hat{y}(i+k+j|\,i) - y^Z(i+k+j)\right]^2 + \boldsymbol{u}^T Q \boldsymbol{u} \tag{5.14}$$

where

$\hat{y}(i+k+j|i)$ – predicted value of the plant output, at the $(i+k+j)$th instant, predicted at the ith instant;

H – plant output prediction horizon;

$\boldsymbol{u} = [u(i)\, u(i+1) \ldots u(i+H)]$ – vector of plant control signal values;

Q – plant control signal weighting matrix.

By solving the stated minimization problem, the following control algorithm is obtained:

$$u(i) = -[1\ 0\ \ldots\ 0](\boldsymbol{G}^T\boldsymbol{G}+\boldsymbol{Q})^{-1}\boldsymbol{G}^T\boldsymbol{r} \tag{5.15}$$

where

$$\boldsymbol{G} = \begin{bmatrix} g_0 & & & 0 \\ g_1 & g_0 & & \\ \vdots & \vdots & \ddots & \\ g_H & g_{H-1} & \cdots & g_0 \end{bmatrix} \tag{5.16}$$

$$\boldsymbol{r}^T = [r_0\, r_1 \ldots r_H] \tag{5.17}$$

$$r_s = \sum_{j=s+1}^{dB} \hat{g}_j\, u(i-j+s) - \beta^{s+1} y^M(i+k-1)$$
$$-(1-\beta^{s+1})\left[y^M(i) - y(i) + w(i)\right] \tag{5.18}$$

with $g(i)$ being the ith element of the discrete-time pulse response of the plant, and $y^M(i)$ the plant model output (in the form of a finite approximation of the plant weighting function):

$$y^M(i) = \sum_{j=0}^{dB} \hat{g}_j\, u(i-k-j) \tag{5.19}$$

Assuming additionally that, beginning from the $(i+k+L)$th time instant, the value of the control signal should be 0 (with L being the control signal prediction horizon [7]), the following version of the control algorithm is obtained:

$$u(i) = -[1\ 0\ \ldots\ 0](\boldsymbol{G}_L^T\boldsymbol{G}_L+\boldsymbol{Q})^{-1}\boldsymbol{G}_L^T\boldsymbol{r} \tag{5.20}$$

where

$$\boldsymbol{G}_L = \begin{bmatrix} g_0 & & & 0 \\ g_1 & g_0 & & \\ \vdots & \vdots & \ddots & \\ g_L & g_{L-1} & \cdots & g_0 \\ \vdots & \vdots & & \vdots \\ g_H & g_{H-1} & \cdots & g_{H-L} \end{bmatrix} \tag{5.21}$$

5.2.5 GPC algorithm

During GPC algorithm derivation we assume that the plant dynamics may be described with the standard ARIMAX model, with the disturbance channel polynomial $C(z^{-1})$ being equal to 1 and $D(z^{-1}) = \nabla = 1 - z^{-1}$.

The aim of GPC algorithm is to minimize the performance index:

$$J = \boldsymbol{e}^T \boldsymbol{e} + q_0 \boldsymbol{\nabla u}^T \boldsymbol{\nabla u} \tag{5.22}$$

where $\boldsymbol{e}$ is the vector of control error defined as:

$$\boldsymbol{e}^T = \left[w(i+k) - \hat{y}(i+k|\,i), \ldots, w(i+k+H) - \hat{y}(i+k+H\,|\,i)\right] \tag{5.23}$$

with $\hat{y}(i+j|\,i)$ the optimal predicted value of the plant output at the $(i+j)$th time instant predicted from the ith, H the plant output prediction horizon. q_0 is the weighting coefficient of the vector of plant control signal values in the minimized performance index.

Minimization of performance index (5.22) leads to the long-range predictive control algorithm:

$$\nabla u(i) = [1\ 0\ \ldots\ 0](\boldsymbol{H}^T \boldsymbol{H} + q_0 \boldsymbol{I})^{-1} \boldsymbol{H}^T (\boldsymbol{w} - \boldsymbol{r}) \tag{5.24}$$

where

$$\boldsymbol{w}^T = \left[w(i+k) \ldots w(i+k+H)\right] \tag{5.25}$$

is the vector of set point values, and

$$\hat{\boldsymbol{y}} = \boldsymbol{H \nabla u} + \boldsymbol{r} \tag{5.26}$$

$$\boldsymbol{H} = \begin{bmatrix} h_0^k & & & 0 \\ h_1^{k+1} & h_0^{k+1} & & \\ \vdots & \vdots & \ddots & \\ h_H^{k+H} & h_{H-1}^{k+H} & \ldots & h_0^{k+H} \end{bmatrix} \tag{5.27}$$

$$\boldsymbol{\nabla u}^T = \left[\nabla u(i) \ldots \nabla u(i+H)\right] \tag{5.28}$$

$$\boldsymbol{r}^T = [r_0\, r_1 \ldots r_H] \tag{5.29}$$

$$r_s = \sum_{j=0}^{dA} g_j^{k+s} y(i-j) + \sum_{j=1}^{dB+k-1} h_{s+j}^{k+s} \left[u(i-j) - u(i-j-1)\right] \tag{5.30}$$

If we assume that starting from time instant $i+k+L$ the plant control signal u should be equal to 0, with L known as the control signal prediction horizon, we obtain the following simplified version of the GPC algorithm:

$$\nabla u(i) = [1\ 0\ \ldots\ 0](\boldsymbol{H}_L^T \boldsymbol{H}_L + q_0 \boldsymbol{I})^{-1} \boldsymbol{H}_L^T (\boldsymbol{w} - \boldsymbol{r}) \tag{5.31}$$

where

$$\boldsymbol{H}_L = \begin{bmatrix} h_0^k & & & 0 \\ h_1^{k+1} & h_0^{k+1} & & \\ \vdots & \vdots & \ddots & \\ h_L^{k+L} & h_{L-1}^{k+L} & \cdots & h_0^{k+L} \\ \vdots & \vdots & & \vdots \\ h_H^{k+H} & h_{H-1}^{k+H} & \cdots & h_{H-L}^{k+H} \end{bmatrix} \tag{5.32}$$

5.2.6 Adaptive PID control with non-parametric identification

There are several methods for deriving the parameters of a PID controller on the basis of the controlled plant frequency response parameters. These methods include the Ziegler–Nichols frequency response method, the dominant poles design and the M_{max} method. All of them are based on the determination of some points of the controlled plant frequency response, the plant critical gain and oscillation frequency being the most popular and easiest parameters to obtain. One of the most popular methods of approximate identification of the plant frequency response is based on the concept of closing the control loop in the tuning phase with a non-linear element of relay type, preferably with some hysteresis. During the initial phase of control, the relay in the feedback channel ensures that after some time oscillations will appear in the control system. The oscillations are continuously monitored and when their period seems to be constant we can measure it along with the oscillation amplitude. Afterwards the corresponding fundamental component is calculated. On the basis of such a modified Ziegler–Nichols experiment, the user knows all the data necessary to calculate the values of the PID controller parameters using the Ziegler–Nichols formulae. If we use a relay with hysteresis or any other more involved non-linear element in the feedback channel, we can gather more information about the plant frequency response parameters and obviously we can use more involved design procedures.

5.3 IDENTIFICATION IN ADAPTIVE SYSTEMS

In an adaptive control system, identification procedures are used to obtain the estimates of the parameters of the weighting function, transfer function or prediction model of the plant. The estimates of the prediction model parameters are either directly used in the adaptive controller equation, or, e.g. in the case of a NMP plant, the estimates are further processed to eliminate the influence of the NMP part of the plant model, $B^-(z^{-1})$.

The following three estimation schemes are commonly used in adaptive control systems:

- recursive least squares (RLS),
- recursive least mean squares (LMS),
- recursive prediction error minimization (RPE).

The RLS and LMS procedures, which are the basic ones, will be described briefly below. The RPE method is usually implemented following its extensive analysis in [19].

5.3.1 Recursive least squares

The RLS method could be crudely implemented on the basis of the recursive equations:

$$\hat{\boldsymbol{\theta}}(i) = \hat{\boldsymbol{\theta}}(i-1) + \boldsymbol{k}(i)\left[\Psi(i) - \boldsymbol{\varphi}^T(i-K)\,\hat{\boldsymbol{\theta}}(i-1)\right] \tag{5.33}$$

$$\boldsymbol{k}(i) = \frac{\boldsymbol{P}(i-1)\boldsymbol{\varphi}(i-K)}{\alpha + \boldsymbol{\varphi}^T(i-K)\boldsymbol{P}(i-1)\boldsymbol{\varphi}(i-K)} \tag{5.34}$$

$$\boldsymbol{P}(i) = \frac{1}{\alpha}\left[\boldsymbol{P}(i-1) - \frac{\boldsymbol{P}(i-1)\boldsymbol{\varphi}(i-K)\boldsymbol{\varphi}^T(i-K)\boldsymbol{P}(i-1)}{\alpha + \boldsymbol{\varphi}^T(i-K)\boldsymbol{P}(i-1)\boldsymbol{\varphi}(i-K)}\right] \tag{5.35}$$

where

$\hat{\boldsymbol{\theta}}(i)$ – model parameter estimates vector, in the ith identification step,
$\Psi(i)$ – model output variable value, in the ith identification step,
$\boldsymbol{\varphi}(i)$ – model regressor vector, in the ith identification step,
K – prediction horizon (time delay) of the model,
$\boldsymbol{k}(i)$ – identification gain vector, in the ith identification step,
$\boldsymbol{P}(i)$ – identification matrix (proportional to covariance of $\hat{\boldsymbol{\theta}}(i)$), in the ith identification step,
α – forgetting factor.

Initial values $\hat{\boldsymbol{\theta}}(0)$ of the estimates are usually chosen as being equal to 0. The initial identification matrix $\boldsymbol{P}(0)$ is usually a diagonal matrix with large coefficients on its main diagonal:

$$\boldsymbol{P}(0) = \boldsymbol{P}_0\,\boldsymbol{I}\,, \qquad \boldsymbol{P}_0 \gg 1 \tag{5.36}$$

If the forgetting factor in the RLS method is equal to 1, then the elements of the identification matrix $\boldsymbol{P}(i)$ decrease with the identification iterations, preventing rapid changes in the estimates after some iterations of the identification process. The possibility of changes in the estimates may be measured by the scalar gain coefficient

$$k_w = \boldsymbol{\varphi}^T(i-K)\boldsymbol{P}(i)\boldsymbol{\varphi}(i-K) \tag{5.37}$$

Rapid changes of the model parameter estimates are possible if either of the following modifications is used:

(a) covariance ($\boldsymbol{P}$ matrix) resetting, periodically increasing the values of all $\boldsymbol{P}(i)$ matrix elements by such a factor that the largest element after the increase is equal to $\boldsymbol{P}_0$ (or $\boldsymbol{P}_0/\gamma$, where γ is some scaling factor); such a mechanism makes it possible to increase the absolute values of the $\boldsymbol{P}$ matrix elements and the value of the scalar gain coefficient k_w, thus allowing fast (and effective) adaptation in the case of time-varying plant behaviour;

(b) using a forgetting factor $\alpha < 1$, causing the $\boldsymbol{P}(i)$ matrix elements calculated from (5.35) not to decrease much during the identification process.

Using a forgetting factor $\alpha < 1$ (if the controlled plant is not excited enough) causes the $\boldsymbol{\varphi}(i-K)$ vector elements to be small and leads to constant increase of the $\boldsymbol{P}(i)$ matrix elements and the gain vector $\boldsymbol{k}(i)$. As a result, when a large identification (prediction) error happens (e.g. because of a sudden, large disturbance) the model parameter estimates $\hat{\boldsymbol{\theta}}(i)$ change quite rapidly, and disturb the process of identification. The effect may be especially disastrous in the direct adaptive control system, in which the estimates are directly used as controller equation coefficients. Methods are described in the literature to avoid such unpleasant identification/control properties. For example, in [1, 10, 15] an estimator information measure is proposed, calculated as the weighted sum of the identification errors:

$$\Sigma(i) = \alpha(i)\,\Sigma(i-1) + \left[1-\boldsymbol{\varphi}^T(i-K)\,\boldsymbol{k}(i)\right]\left[\Psi(i)-\boldsymbol{\varphi}^T(i-K)\,\hat{\boldsymbol{\theta}}(i-1)\right]^2 \qquad (5.38)$$

The forgetting factor $\alpha(i)$ is adjusted at each identification step to keep the estimator information measure (5.38) constant, i.e. the user wants

$$\Sigma(i) = \Sigma(i-1) = \cdots = \Sigma(0) \qquad (5.39)$$

where $\Sigma(0)$ (the initial value of the $\Sigma(i)$) may be used for tuning the forgetting factor adjustment, in the presence of time-varying plant behaviour.

Under the above assumptions, the forgetting factor for the ith step of identification is calculated by

$$\alpha(i) = 1 - \left[1-\boldsymbol{\varphi}^T(i-K)\,\boldsymbol{k}(i)\right]\left[\Psi(i)-\boldsymbol{\varphi}^T(i-K)\,\hat{\boldsymbol{\theta}}(i-1)\right]^2/\Sigma(0) \qquad (5.40)$$

Another method used to ensure good properties of the RLS estimation procedure for lengthy experiments is to keep the trace of $\boldsymbol{P}(i)$ constant, and to introduce an identification error dead zone: if the identification error remains small enough, the update procedures are not activated at all. An adequate sequence of recursive formulae for this version of the RLS method is

$$\hat{\boldsymbol{\theta}}(i) = \hat{\boldsymbol{\theta}}(i-1) + a(i)\boldsymbol{k}(i)\left[\Psi(i)-\boldsymbol{\varphi}^T(i-K)\,\hat{\boldsymbol{\theta}}(i-1)\right] \qquad (5.41)$$

$$\boldsymbol{k}(i) = \frac{\boldsymbol{P}(i-1)\boldsymbol{\varphi}(i-K)}{1+\boldsymbol{\varphi}^T(i-K)\boldsymbol{P}(i-1)\boldsymbol{\varphi}(i-K)+\bar{c}\boldsymbol{\varphi}^T(i-K)\boldsymbol{\varphi}} \qquad (5.42)$$

$$\bar{\boldsymbol{P}}(i) = \boldsymbol{P}(i-1) - a(i)\frac{\boldsymbol{P}(i-1)\boldsymbol{\varphi}(i-K)\boldsymbol{\varphi}^T(i-K)\boldsymbol{P}(i-1)}{1+\boldsymbol{\varphi}^T(i-K)\boldsymbol{P}(i-1)\boldsymbol{\varphi}(i-K)+\bar{c}\boldsymbol{\varphi}^T(i-K)\boldsymbol{\varphi}} \qquad (5.43)$$

$$\boldsymbol{P}(i) = C_1\frac{\bar{\boldsymbol{P}}(i)}{\mathrm{tr}(\bar{\boldsymbol{P}}(i))} + C_2\boldsymbol{I} \qquad (5.44)$$

$$a(i) = \begin{cases} \bar{a} & \text{for} \quad |\Psi(i)-\boldsymbol{\varphi}^T(i-K)\,\hat{\boldsymbol{\theta}}(i-1)| > 2\delta \\ 0 & \text{if the above condition does not hold} \end{cases}$$

where

C_1 – specified value of P matrix trace,
C_2 – P matrix diagonalization factor,
δ – identification error dead zone,
$\bar{a}$ – identification error weighting coefficient,
$\bar{c}$ – a factor used to maintain sufficiently large values of the P matrix elements.

A so-called improved least squares estimation method is also used. The convergence, stability and speed properties of this version are achieved by normalization of the regressor vector, scaling of the $\boldsymbol{P}(i)$ matrix to ensure the minimum condition number, adjusting the forgetting factor, and avoiding updating the $\boldsymbol{P}(i)$ matrix and $\hat{\boldsymbol{\theta}}(i)$ vector during periods of low excitation. The recursive calculations are as follows:

1. Normalization of the regressor vector:

$$n(i) = \max(1, \|\varphi(\boldsymbol{i} - \boldsymbol{K})\|) \tag{5.45}$$

$$\Psi_n(i) = \Psi(i)/n(i) \tag{5.46}$$

$$\varphi_n(i-K) = \varphi(i-K)/n(i) \tag{5.47}$$

2. Calculating the forgetting factor value to stabilize $P(i)$ matrix trace:

$$r(i) = 1 + \varphi_n^T(i-K)\boldsymbol{P}(i-1)\varphi_n(i-K) \tag{5.48}$$

$$\alpha(i) = 1 - 0.5\left[r(i) - \left\{r(i)^2 - 4\frac{\|\boldsymbol{P}(i-1)\varphi_n(i-K)\|^2}{\mathrm{tr}\,\boldsymbol{P}(i-1)}\right\}^{\frac{1}{2}}\right] \tag{5.49}$$

3. Testing the excitation properties:

$$N(i) = \|\boldsymbol{P}(i-1)\varphi_n(i-K)\| \tag{5.50}$$

If $N(i) < N_{min}$ (low excitation), then the calculation proceeds to 7.

4. Calculation of new values of $\boldsymbol{P}$:

$$\bar{\boldsymbol{P}}(i) = \frac{1}{\alpha(i)}\left[\boldsymbol{P}(i-1) - \frac{\boldsymbol{P}(i-1)\varphi_n(i-K)\varphi_n^T(i-K)\boldsymbol{P}(i-1)}{\alpha(i) + \varphi_n^T(i-K)\boldsymbol{P}(i-1)\varphi_n(i-K)}\right] \tag{5.51}$$

$$\bar{\boldsymbol{P}}^{-1}(i) = \alpha(i)\boldsymbol{P}^{-1}(i-1) + \varphi_n(i-K)\varphi_n^T(i-K) \tag{5.52}$$

$$C\{\bar{\boldsymbol{P}}\} = \|\bar{\boldsymbol{P}}\|_\infty \|\bar{\boldsymbol{P}}^{-1}\|_\infty \tag{5.53}$$

If $C\{\bar{\boldsymbol{P}}\} > C_{max}$ (poorly conditioned $\boldsymbol{P}$), then the calculation proceeds to 6.

5. Updating the identified parameter vector and $\boldsymbol{P}$:

$$\boldsymbol{P}(i) = \bar{\boldsymbol{P}}(i), \quad \boldsymbol{P}^{-1}(i) = \bar{\boldsymbol{P}}^{-1}(i), \quad \boldsymbol{S}(i) = \boldsymbol{S}(i-1) \tag{5.54}$$

$$\hat{\boldsymbol{\theta}}(i) = \hat{\boldsymbol{\theta}}(i-1) + \boldsymbol{S}^{-1}(i-1)\boldsymbol{P}(i)\varphi_n(i-K) \times \left[\Psi_n(i) - \varphi_n^T(i-K)\hat{\boldsymbol{\theta}}(i-1)\right] \tag{5.55}$$

The calculation proceeds to 9.

6. Calculating the elements of the scaling matrices. $\bar{\boldsymbol{S}}$ is calculated to minimize the condition number $\bar{\boldsymbol{S}}\bar{\boldsymbol{P}}(i)$, then $\bar{\boldsymbol{P}}(i)$ and $\bar{\boldsymbol{P}}^{-1}(i)$ are updated:

$$\bar{\boldsymbol{P}}(i) = \bar{\boldsymbol{S}}\bar{\boldsymbol{P}}(i)\bar{\boldsymbol{S}}, \quad \bar{\boldsymbol{P}}^{-1}(i) = \bar{\boldsymbol{S}}^{-1}\bar{\boldsymbol{P}}^{-1}(i)\bar{\boldsymbol{S}}^{-1} \tag{5.56}$$

   If $C\{\bar{\boldsymbol{P}}\} <= C_{max}$ (sufficiently well conditioned $\boldsymbol{P}$), the calculation proceeds to 8.

7. Stopping the estimation:

$$\boldsymbol{P}(i) = \boldsymbol{P}(i-1), \quad \boldsymbol{P}^{-1}(i) = \boldsymbol{P}^{-1}(i-1), \quad \boldsymbol{S}(i) = \boldsymbol{S}(i-1) \tag{5.57}$$

$$\hat{\boldsymbol{\theta}}(i) = \hat{\boldsymbol{\theta}}(i-1) \tag{5.58}$$

8. Updating the identified parameter values and $\boldsymbol{P}$:

$$\boldsymbol{P}(i) = \bar{\boldsymbol{P}}(i), \quad \boldsymbol{P}^{-1}(i) = \bar{\boldsymbol{P}}^{-1}(i) \tag{5.59}$$

$$\boldsymbol{S}(i) = \boldsymbol{S}(i-1)\bar{\boldsymbol{S}}, \quad \varphi_n(i-K) = \bar{\boldsymbol{S}}^{-1}\varphi_n(i-K). \tag{5.60}$$

$$\hat{\boldsymbol{\theta}}(i) = \hat{\boldsymbol{\theta}}(i-1) + \boldsymbol{S}^{-1}(i)\boldsymbol{P}(i)\varphi_n(i-K)\left[\Psi_n(i) - \varphi_n^T(i-K)\hat{\boldsymbol{\theta}}(i-1)\right] \tag{5.61}$$

9. Constraining the parameter estimates to some chosen neighbourhood of the initial guesses $\hat{\boldsymbol{\theta}}(0)$. If $\boldsymbol{R}$ denotes the radius of a sphere in $\hat{\boldsymbol{\theta}}$ space, in which all parameter estimates have to lie, then the estimates may be constrained:

$$\hat{\boldsymbol{\theta}}_D(i) = \hat{\boldsymbol{\theta}}(i) - \hat{\boldsymbol{\theta}}(0) \tag{5.62}$$

$$\hat{\boldsymbol{\theta}}(i) = \hat{\boldsymbol{\theta}}(0) + \min\left(1, \frac{R}{||\hat{\boldsymbol{\theta}}_D(i)||}\right)\hat{\boldsymbol{\theta}}_D(i) \tag{5.63}$$

In the improved least squares method, the user may choose the minimum value of the forgetting factor calculated by (5.49), the minimum value N_{min} of the norm in (5.50), the maximum value C_{max} of the condition factor, and the maximum radius R in (5.63).

5.3.2 Recursive least mean squares

The recursive least mean squares (LMS) method is much simpler and faster than recursive least squares. The LMS method may be described by

$$\begin{aligned}\hat{\boldsymbol{\theta}}(i) &= \hat{\boldsymbol{\theta}}(i-1) + \varphi(i-K)\frac{\gamma_K}{\alpha_K + \varphi^T(i-K)\varphi(i-K)} \\ &\quad \times \left[\Psi(i) - \varphi^T(i-K)\,\hat{\boldsymbol{\theta}}(i-1)\right]\end{aligned} \tag{5.64}$$

where

$\hat{\theta}(i)$ – parameter estimates vector,
$\Psi(i)$ – model output,
$\varphi(i)$ – regressor vector,
K – prediction horizon (time delay),
γ_K – gain coefficient,
α_K – forgetting factor.

While using the recursive LMS estimation method, the user may adjust the two important parameters:

α_K – the forgetting factor, usually restricted to the interval $(0.0, 10.0 >$;

γ_K – the gain coefficient, usually in the interval $(0.0, 2.0 >$.

Increasing α_K (or decreasing γ_K) worsens the convergence properties of the LMS method, but prevents large variations in the parameter estimates (especially dangerous in adaptive control).

5.4 IDENTIFICATION MODEL STRUCTURE

In indirect adaptive control we must identify the parameters of a "classic" model of the plant, for example the weighting function or transfer-function model. The choice of model structure and identification scheme has been examined in detail by many authors, see e.g. [19].

Direct adaptive control needs a plant model in which the R, S and T polynomials and the h term are present, constituting the controller equation. This is the case for a *prediction model* of the plant, which is also linear in the parameters, making it possible to identify them using least squares.

The way in which the prediction model parameters are identified depends on the minimum phase property of the plant, or on how information about the plant being nonminimum phase is incorporated.

5.4.1 Minimum phase plant

If the plant is minimum phase, the only unknown parameters in the prediction model are the resulting control algorithm parameters (with respect to which the model is linear). This means that the user may identify the parameters, and use them directly as the controller coefficients. This way of implementing the minimum-variance and pole/zero-placement control algorithms is conceptually easy and numerically effective. Such a self-tuning control system is valid only for a minimum phase plant.

5.4.2 Non-minimum phase plant

If the plant is non-minimum phase, some elements in the corresponding prediction model will be related to this (specifically the B^- polynomial). An example (similar to the one described by (5.7)) is:

$$\phi(i) - RB^-u(i-k) - SB^-y(i-k) + TB^-w(i-k) - hB^-(1) = \epsilon(i) \quad (5.65)$$

where $\epsilon(i)$ is the model error.

The estimation of such model-coefficient values should be effective and result in the controller parameters (R, S and T polynomials and the h term). This prediction model structure suggests at least two ways to achieve the stated aim:

- Identify jointly the parameters RB^-, SB^-, TB^- and $hB^-(1)$ in a prediction model

$$\phi(i) - Lu(i-k) - My(i-k) + Nw(i-k) - h_{B^-} = \epsilon(i) \tag{5.66}$$

  then conclude the controller parameters, R, S, T and h, by dividing L, M and N by B^-, and h_{B^-} by $B^-(1)$.

- At the beginning of each identification step, filter the regression signals $u(i-k)$, $y(i-k)$ and $w(i-k)$ with B^-, and the constant 1 with $B^-(1)$, leading to the prediction model:

$$\phi(i) - Ru_{B^-}(i-k) - Sy_{B^-}(i-k) + Tw_{B^-}(i-k) - h1_{B^-} = \epsilon(i) \tag{5.67}$$

  Such a model can be directly used to identify the control parameters: R, S, T and h.

These solutions assume that $B^-(z^{-1})$ is known. This may be so if
either

- the user pretends to know B^-'s degree and coefficients on the basis of his/her plant knowledge (maybe as a result of previous identification);

or

- B^- is identified jointly with the parameters of the controller; the possible solution is to use the prediction model:

$$\phi(i) - Lu(i-k) - Sy_{B^-}(i-k) + Tw_{B^-}(i-k) - h \cdot 1_{B^-} = \epsilon(i) \tag{5.68}$$

  At the end of each identification step, the unstable parts of L are grouped and processed further as B^-, the rest of the L estimate being treated as a controller polynomial R. In the next identification step, the estimate of B^- serves to filter the regression variables $y(i-k)$ and $w(i-k)$ and the constant 1. In the first identification step we may assume that $B^-(z^{-1}) = 1$.

In specific cases some factors of the R, S and T polynomials may be known and thus need not be identified. If we denote such terms as R_k, S_k and T_k and with R_u, S_u and T_u unknown, then during adaptive control the known factors of the controller polynomials should be used for filtering the appropriate regressor signals before the estimation step, while the unknown factors have to be identified (maybe jointly with the B^- polynomial).

5.5 ESTIMATION OF NON-STATIONARY PLANT PARAMETERS

As is common practice in adaptive control and in CACSD-based adaptive control simulation, the user is free to choose among at least three estimation schemes:

(a) recursive least squares (RLS), in its standard version, the constant tr P version, and improved version,

(b) recursive least mean squares (LMS),

(c) recursive prediction error minimization (RPE).

The LMS method is considerably faster than both RLS and RPE. However, for many adaptive control systems, the RLS method is considerably better for parameter-estimate convergence rate and bias.

The basic parameter of weighted RLS is the forgetting factor α. The user should choose α equal to 1 for a plant assumed to be stationary, or $0 \ll \alpha < 1$ for a plant assumed non-stationary. Any constant α may cause slow convergence of the parameter estimates, or periodic large changes of their values. Usually the forgetting factor may be continuously varied according to the weighted sum of squares of the model prediction error [10]. The user may choose the lower bound of the adjusted forgetting factor and the gain of the adjustment procedure. Similar good properties may be obtained when using the constant tr $\boldsymbol{P}$ version of RLS [5], or the improved (or robust) least squares method [25]. These methods ensure fast adaptation after a change of plant parameters and there is no danger related to low excitation of the plant (no "bursts" in the estimates).

The initial values of the main diagonal elements of $\boldsymbol{P}$ are also important in the RLS method. Initial values that are too large may cause temporary destabilization of the adaptive control system. On the other hand, small initial values may cause slow convergence of the parameter estimates.

RLS is especially vigorous in the first few iteration steps. This suggests restarting periodically with the initial $\boldsymbol{P}_0$ and the last vector of the parameter estimates (covariance resetting). Covariance resetting may be also requested if the user considers that the dynamics of the system have changed. The covariance resetting procedure makes RLS much faster in tracking changing parameters than the standard weighted RLS method with forgetting factor less than 1. On the other hand, covariance resetting may cause violent changes in the parameter estimates. It should be stressed that both the constant tr $\boldsymbol{P}$ version of RLS identification and improved least squares need no periodical resetting of the covariance, while achieving fast adaptation in the presence of non-stationary plant behaviour.

5.6 MULTIVARIABLE MINIMUM-VARIANCE ADAPTIVE CONTROL

Multivariable versions of the minimum-variance control schemes have appeared in the literature for more than 15 years now, starting from the early works of Borisson and Koivo [6, 17]. Further work and extensions have been presented by, among others, Dugard and Scattolini *et al.* [8, 23, 22].

It is supposed that the multivariable plant to be controlled is described by the linear model:

$$\boldsymbol{A}(z^{-1})\boldsymbol{y}(i) = \boldsymbol{B}(z^{-1})\boldsymbol{u}(i-k) + \boldsymbol{C}(z^{-1})\boldsymbol{e}(i) + \boldsymbol{d} \tag{5.69}$$

with $\boldsymbol{u}(i)$ a p-dimensional control signal, $\boldsymbol{y}(i)$ a p-dimensional output (controlled) signal, $\boldsymbol{e}(i)$ a p-dimensional white noise sequence with diagonal covariance matrix, and $\boldsymbol{d}$ a p-dimensional constant disturbance (bias) vector. $\boldsymbol{A}(z^{-1})$, $\boldsymbol{B}(z^{-1})$ and $\boldsymbol{C}(z^{-1})$ are polynomial matrices, with $\boldsymbol{A}$ being monic. Argument z^{-1} will be omitted from here on.

It is assumed that the plant is minimum phase in the multivariable sense, i.e. that $\boldsymbol{B}$ is stable in such sense. It is also assumed that $\boldsymbol{C}$ is stable.

The performance index to be minimized in the multivariable generalized minimum-variance scheme is

$$\min_{\boldsymbol{u}(i)} \left[\mathrm{E}\left\{ ||\boldsymbol{y}(i+k) - \boldsymbol{P}^{-1}\boldsymbol{V}\boldsymbol{w}(i)||^2 + ||\boldsymbol{Q}\boldsymbol{u}(i)||^2 \right\} \right] \tag{5.70}$$

where $\boldsymbol{w}(i)$ denotes a p-dimensional set point vector, and $\boldsymbol{P}$, $\boldsymbol{V}$ and $\boldsymbol{Q}$ are $p \times p$ matrices in the z^{-1} operator.

The control algorithm resulting from this problem statement is defined by

$$\left[\tilde{\boldsymbol{F}}\boldsymbol{B} + \tilde{\boldsymbol{C}}\boldsymbol{B}_0^{-T}\boldsymbol{Q}_0^T\boldsymbol{Q}\right] \boldsymbol{u}(i) + \tilde{\boldsymbol{G}}\boldsymbol{y}(i) - \tilde{\boldsymbol{C}}\boldsymbol{w}^*(i) + \tilde{\boldsymbol{F}}(1)\boldsymbol{d} = \mathbf{o} \tag{5.71}$$

$$\boldsymbol{C} = \boldsymbol{A}\boldsymbol{F} + z^{-k}\boldsymbol{G} \tag{5.72}$$

$$\tilde{\boldsymbol{C}}\boldsymbol{F} = \tilde{\boldsymbol{F}}\boldsymbol{C} \tag{5.73}$$

$$\tilde{\boldsymbol{F}}\boldsymbol{G} = \tilde{\boldsymbol{G}}\boldsymbol{F} \tag{5.74}$$

$$\boldsymbol{P}\boldsymbol{w}^*(i) = \boldsymbol{V}\boldsymbol{w}(i) \tag{5.75}$$

with $\boldsymbol{F}$ and $\tilde{\boldsymbol{F}}$ $p \times p$ polynomial matrices of $k-1$ degree, and $\boldsymbol{G}$ and $\tilde{\boldsymbol{G}}$ also $p \times p$ polynomial matrices, of degree sufficient for the identity (5.72) to be solvable.

It is apparent that the control equation may be rewritten in the following, commonly adopted form:

$$\boldsymbol{R}\boldsymbol{u}(i) + \boldsymbol{S}\boldsymbol{y}(i) - \boldsymbol{T}\boldsymbol{w}^*(i) + \boldsymbol{\delta} = \mathbf{o} \tag{5.76}$$

Thus, the control vector could be calculated using

$$\begin{aligned} \boldsymbol{u}_{opt}(i) = -\tilde{\boldsymbol{R}}_0^{-1} \Bigg[& \sum_{j=1}^{n\boldsymbol{R}} \tilde{\boldsymbol{R}}_j \boldsymbol{u}_{opt}(i-j) + \sum_{j=0}^{n\boldsymbol{S}} \tilde{\boldsymbol{S}}_j \boldsymbol{y}(i-j) \\ & - \boldsymbol{w}^*(i) \sum_{j=1}^{n\boldsymbol{T}} \tilde{\boldsymbol{T}}_j \boldsymbol{w}^*(i-j) + \boldsymbol{\delta} \Bigg] \end{aligned} \tag{5.77}$$

with

$$\boldsymbol{R} = \tilde{\boldsymbol{F}}\boldsymbol{B} + \tilde{\boldsymbol{C}}\boldsymbol{B}_0^{-T}\boldsymbol{Q}_0^T\boldsymbol{Q} \tag{5.78}$$

$$\boldsymbol{S} = \tilde{\boldsymbol{G}} \tag{5.79}$$

$$\boldsymbol{T} = \tilde{\boldsymbol{C}} \tag{5.80}$$

$$\boldsymbol{\delta} = \tilde{\boldsymbol{F}}(1)\boldsymbol{d} \tag{5.81}$$

It should be observed that, in the case of no control signal weighting (i.e. $Q = 0$), the assumption concerning the minimum phase property of the controlled plant becomes active.

It should be observed also that the plant model may be transformed to include the polynomial matrices appearing in the multivariable minimum-variance control algorithm just given. The appropriate version of the plant model is

$$\begin{aligned} \boldsymbol{\Psi}(i+k) - \left[\tilde{\boldsymbol{F}}\boldsymbol{B} + \tilde{\boldsymbol{C}}\boldsymbol{B}_{0}^{-T}\boldsymbol{Q}_{0}^{T}\boldsymbol{Q}\right]\boldsymbol{u}(i) - \tilde{\boldsymbol{G}}\boldsymbol{y}(i) + \tilde{\boldsymbol{C}}\boldsymbol{w}^{*}(i) - \boldsymbol{q} \\ = \left(\tilde{\boldsymbol{C}} - \mathbf{1}\right)\left[\boldsymbol{F}\boldsymbol{e}(i+k) - \boldsymbol{\Psi}(i+k)\right] + \boldsymbol{F}\boldsymbol{e}(i+k) \end{aligned} \tag{5.82}$$

with

$$\boldsymbol{\Psi}(i+k) = \boldsymbol{y}(i+k) - \boldsymbol{w}^{*}(i) + \boldsymbol{B}_{0}^{-T}\boldsymbol{Q}_{0}^{T}\boldsymbol{Q}\boldsymbol{u}(i) \tag{5.83}$$

The model (5.82) could then be rewritten in the *prediction model* form (regressor model form):

$$\boldsymbol{\Psi}(i) - \boldsymbol{R}\boldsymbol{u}(i-k) - \boldsymbol{S}\boldsymbol{y}(i-k) + \boldsymbol{T}\boldsymbol{w}^{*}(i-k) - \boldsymbol{q} = \boldsymbol{\epsilon}(i) \tag{5.84}$$

or, assuming that some factors of the controller polynomial matrices are known, in the form:

$$\boldsymbol{\Psi}(i) - \boldsymbol{R}_u\boldsymbol{R}_k\boldsymbol{u}(i-k) - \boldsymbol{S}_u\boldsymbol{S}_k\boldsymbol{y}(i-k) + \boldsymbol{T}_u\boldsymbol{T}_k\boldsymbol{w}^{*}(i-k) - \boldsymbol{q} = \boldsymbol{\epsilon}(i) \tag{5.85}$$

in which only the unknown factors and vectors $\boldsymbol{R}_u$, $\boldsymbol{S}_u$, $\boldsymbol{T}_u$ and $\boldsymbol{q}$ need be estimated, if this prediction model is used in the multivariable adaptive control framework. The controller polynomial matrices appearing in (5.85) are of special structure and different dimensions compared with $\boldsymbol{R}$, $\boldsymbol{S}$ and $\boldsymbol{T}$, e.g. $\boldsymbol{R}_u$ and $\boldsymbol{R}_k$ are defined as

$$\boldsymbol{R}_u = \begin{bmatrix} \boldsymbol{R}_{u,11} & & \mathbf{0} & \cdots & \boldsymbol{R}_{u,1p} & & \mathbf{0} \\ & \ddots & & \cdots & & \ddots & \\ \mathbf{0} & & \boldsymbol{R}_{u,p1} & \cdots & \mathbf{0} & & \boldsymbol{R}_{u,pp} \end{bmatrix} \tag{5.86}$$

$$\boldsymbol{R}_k = \begin{bmatrix} \boldsymbol{R}_{k,11} & & \mathbf{0} \\ & \ddots & \\ \boldsymbol{R}_{k,p1} & & \mathbf{0} \\ \vdots & \ddots & \vdots \\ \mathbf{0} & & \boldsymbol{R}_{k,1p} \\ & \ddots & \\ \mathbf{0} & & \boldsymbol{R}_{k,pp} \end{bmatrix} \tag{5.87}$$

Prediction model (5.85) together with the controller equation (5.77) form the basis of the multivariable adaptive minimum-variance control scheme.

5.7 MULTIVARIABLE PREDICTIVE ADAPTIVE CONTROL

Below, a multivariable case of generalized predictive control will be considered, as a representative of multivariable versions of long-range predictive control algorithms.

Multivariable versions of the GPC control algorithm emerged as natural generalizations of the basic GPC control algorithm, their background being traceable to [9, 11]. The version presented here resembles that presented by Shah [24], as well as similar results obtained by Kinnaert [16] and Gu [13].

It is assumed that the plant to be controlled is described by the multivariable model:

$$\boldsymbol{A}\boldsymbol{\Delta}\boldsymbol{y}(i) = \boldsymbol{B}\boldsymbol{\Delta}\boldsymbol{u}(i-k) + \boldsymbol{e}(i) \tag{5.88}$$

where $\boldsymbol{\Delta}$ denotes the diagonal z^{-1} operator matrix with all diagonal elements difference operators $1 - z^{-1}$, all other elements of the model having been introduced in the previous section.

The performance index to be minimized in step i of the GPC algorithm is

$$J_V(i) = \mathrm{E}\left\{\sum_{j=N_1}^{N_2} \|\boldsymbol{y}(i+j) - \boldsymbol{w}(i+j)\|^2 + \sum_{j=0}^{N_2} \|\boldsymbol{\Lambda}\boldsymbol{\Delta}\boldsymbol{u}(i+j)\|^2\right\} \tag{5.89}$$

where $\boldsymbol{\Lambda}$ is the control weighting matrix, and $\boldsymbol{\Delta}\boldsymbol{u}(i+j)$ vector serves as a vector of variables with respect to which the performance index is minimized.

In practice, the following related performance index will be used in subsequent calculations:

$$J(i) = \sum_{j=N_1}^{N_2} \|\hat{\boldsymbol{y}}(i+j|i) - \hat{\boldsymbol{w}}(i+j|i)\|^2 + \sum_{j=0}^{N_2} \|\boldsymbol{\Lambda}\boldsymbol{\Delta}\boldsymbol{u}(i+j)\|^2 \tag{5.90}$$

where $\hat{\boldsymbol{y}}(i+j)$ denotes the vector of optimal predictions of the controlled variables, assuming at time instant i zero values of all future white noise sequence elements; $\hat{\boldsymbol{w}}(i+j)$ denotes the vector of set point predictions. Two cases could be considered:

- Perfect knowledge of the future set point sequence is available, in which case $\hat{\boldsymbol{w}}(i+j|i)=\boldsymbol{w}(i+j)$.

- No knowledge of the future set point sequence is available, giving $\hat{\boldsymbol{w}}(i+j|i)=\boldsymbol{w}(i)$ as the only reasonable solution.

The following observations, concerning the assumptions just imposed, should be made:

1. Usually, in long-range predictive control, a receding horizon concept is adopted, i.e. the optimization performed at each instant i gives the whole vector of "optimal" control increments, only the first element of which is used in practice, the whole calculation procedure being repeated in each step, producing new, updated values of consecutive $\boldsymbol{u}$ signal increments.

2. Usually we assume that after some specified number of control periods, the increments of the control signal should be equal to zero, resulting in the upper index of the sum relating to $\boldsymbol{u}$ in (5.90) being N_u instead of N_2:

$$J(i) = \sum_{j=N_1}^{N_2} \|\hat{\boldsymbol{y}}(i+j|i) - \hat{\boldsymbol{w}}(i+j|i)\|^2 + \sum_{j=0}^{N_u} \|\boldsymbol{\Lambda}\boldsymbol{\Delta u}(i+j)\|^2 \tag{5.91}$$

This so-called control horizon may in practice be much less than the controlled variable prediction horizon (or simply prediction horizon) N_2, greatly simplifying and accelerating the calculations required.

3. All prediction horizons, N_1, N_2 and N_u, were given as scalars above. However, nothing prevents them from being defined as appropriate p-dimensional vectors, corresponding to the assumed structure of the multivariable plant model. In what follows, the scalar version of the presentation relating to the prediction horizons will be retained, so as not to complicate the notation.

Let us define the following aggregate vectors:

$$\hat{\boldsymbol{y}}^*(i) = \left[\hat{\boldsymbol{y}}(i+N_1|i)^T \hat{\boldsymbol{y}}(i+N_1+1|i)^T \ldots \hat{\boldsymbol{y}}(i+N_2|i)^T\right]^T \tag{5.92}$$

$$\hat{\boldsymbol{w}}^*(i) = \left[\hat{\boldsymbol{w}}(i+N_1|i)^T \hat{\boldsymbol{w}}(i+N_1+1|i)^T \ldots \hat{\boldsymbol{w}}(i+N_2|i)^T\right]^T \tag{5.93}$$

$$\boldsymbol{\Delta u}^*(i) = \left[\boldsymbol{\Delta u}(i)^T \boldsymbol{\Delta u}(i+1)^T \ldots \boldsymbol{\Delta u}(i+N_u)^T\right]^T \tag{5.94}$$

The performance index to be minimized (5.91) may be rewritten using the new notation as

$$J(i) = \left\|\hat{\boldsymbol{y}}^*(i) - \hat{\boldsymbol{w}}^*(i)\right\|^2 + \|\boldsymbol{\Lambda}\boldsymbol{\Delta u}^*(i)\|^2 \tag{5.95}$$

The vector of predicted values of the controlled signal can be shown to be

$$\hat{\boldsymbol{y}}^*(i) = \boldsymbol{H}\boldsymbol{\Delta u}^*(i) + \boldsymbol{r}(i) \tag{5.96}$$

with $\boldsymbol{\Delta u}^*(i)$ defined as in (5.94), i.e. including only future increments of the control vector, while the influence of all past values of the control vector on the vector of predicted values of the controlled signal is grouped in the vector $\boldsymbol{r}(i)$ vector, of dimension $(N_2 - N_1) \times p$. Thus it is clear that the predicted values of the controlled signal $\hat{\boldsymbol{y}}^*(i)$, with the constraint that all the future increments of the control signal are zero ("free" prediction), would be $\boldsymbol{r}(i)$, giving easy on-line calculation of the elements of vector $\boldsymbol{r}(i)$, if the model of the multivariable plant is known.

The $\boldsymbol{H}$ matrix appearing in (5.96) is the plant model step response block matrix:

$$\boldsymbol{H} = \begin{bmatrix} \boldsymbol{h}_{N_1-1} & & & \mathbf{o} \\ \boldsymbol{h}_{N_1} & \boldsymbol{h}_{N_1-1} & & \\ \vdots & \vdots & \ddots & \\ \boldsymbol{h}_{N_u} & \boldsymbol{h}_{N_u-1} & \cdots & \boldsymbol{h}_{\mathrm{o}} \\ \vdots & \vdots & & \vdots \\ \boldsymbol{h}_{N_2-1} & \boldsymbol{h}_{N_2-2} & \cdots & \boldsymbol{h}_{N_2-1-N_u} \end{bmatrix} \tag{5.97}$$

with each $\boldsymbol{h}_i$ element a $p \times p$ matrix of i-th elements of the multivariable plant step response.

Incorporating (5.96) into the performance index (5.91) and completing the minimization results in the final control algorithm:

$$\Delta \boldsymbol{u}^*_{opt}(i) = \left(\boldsymbol{H}^T \boldsymbol{H} + \boldsymbol{\Lambda}\right)^{-1} \boldsymbol{H}^T \left(\boldsymbol{w}^*(i) - \boldsymbol{r}(i)\right) \tag{5.98}$$

from which the required current control increments $\Delta \boldsymbol{u}_{opt}(i)$ may be extracted.

We can conclude that, for adaptive multivariable GPC control, the following steps can be adopted in each iteration of the multivariable plant control:

1. Identify the parameters of the plant model in the form of (5.88).
2. Calculate the estimates of the multivariable plant step response $\boldsymbol{H}$ matrix, in the form defined by (5.97), on the basis of the plant model obtained in step 1.
3. Calculate the estimates of the multivariable plant "free" response prediction vector $\boldsymbol{r}(i)$, as defined by (5.96), also on the basis of the plant model identification.
4. Calculate the optimal control increments using (5.98), and extract the first p elements of the $\Delta \boldsymbol{u}^*(i)$ vector.

5.8 MULTIVARIABLE SYSTEM IDENTIFICATION

System identification methods are inevitably present in the adaptive control systems, and usually affect their behaviour quite substantially. A lot of work in the adaptive control field in practice is devoted to making the underlying identification schemes faster, more reliable and more robust. It is agreed that the multivariable identification case can be even more difficult.

In what follows some concepts concerning multivariable identification methods will be outlined. However, the coverage does not pretend to be either very elaborate or very thorough. The methods given below follow closely the standard ones, which may be found for example in [19], with some numerically oriented enhancements as suggested, for example, in [18].

Basically, we can use two different kinds of multivariable plant model:

- a *plant-oriented model*, also called a transfer-function model (following the SISO case), as described, for example, by (5.88); this kind of model is used in multivariable GPC;
- a *controller-oriented model*, also called a *prediction model*, introduced by (5.82), (5.84) and eventually (5.85); this kind of model is used for multivariable generalized minimum-variance control.

In both cases, the plant model to be identified is of the general form:

$$\boldsymbol{\Psi}(i) - \hat{\boldsymbol{\Theta}}^T \boldsymbol{\Phi}(i) = \boldsymbol{\epsilon}(i) \tag{5.99}$$

with $\boldsymbol{\Psi}(i)$ the output of the model, $\hat{\boldsymbol{\Theta}}^T$ the model parameter matrix, $\boldsymbol{\Phi}(i)$ the regressors vector, and $\boldsymbol{\epsilon}(i)$ the model error vector.

For the plant-oriented model case, $\boldsymbol{\Psi}(i)$ is simply $\boldsymbol{\Delta y}(i)$, as in (5.88), whereas $\boldsymbol{\Phi}(i)$ is formed from the delayed $\boldsymbol{\Delta y}(i)$ and $\boldsymbol{\Delta u}(i)$ signal increment vectors, with the $\boldsymbol{A}$ and $\boldsymbol{B}$ polynomial matrices building $\hat{\boldsymbol{\Theta}}^T$.

Estimation is usually performed through consecutive rows of the (5.99) model, i.e. the p sub-models of the form:

$$\boldsymbol{\Psi}^j(i) - \hat{\boldsymbol{\Theta}}^{j^T} \boldsymbol{\Phi}(i) = \epsilon^j(i) \tag{5.100}$$

with $\boldsymbol{\Psi}^j(i)$ the jth output of the model, $\hat{\boldsymbol{\Theta}}^{j^T}$ the jth row of the parameter estimate matrix $\hat{\boldsymbol{\Theta}}^T$, and $\epsilon^j(i)$ the jth element of the model error vector. Estimation then proceeds through the well-known formulae, e.g. in weighted recursive least squares, the calculations during each step are:

$$\hat{\boldsymbol{\Theta}}^j(i) = \hat{\boldsymbol{\Theta}}^j(i-1) + \boldsymbol{k}(i)\left[\boldsymbol{\Psi}^j(i) - \hat{\boldsymbol{\Theta}}^{j^T}(i-1)\boldsymbol{\Phi}(i)\right] \tag{5.101}$$

$$\boldsymbol{k}(i) = \frac{\boldsymbol{P}(i-1)\boldsymbol{\Phi}(i)}{\alpha + \boldsymbol{\Phi}^T(i)\boldsymbol{P}(i-1)\boldsymbol{\Phi}(i)} \tag{5.102}$$

$$\boldsymbol{P}(i) = \frac{1}{\alpha}\left[\boldsymbol{P}(i-1) - \frac{\boldsymbol{P}(i-1)\boldsymbol{\Phi}(i)\boldsymbol{\Phi}^T(i)\boldsymbol{P}(i-1)}{\alpha + \boldsymbol{\Phi}^T(i)\boldsymbol{P}(i-1)\boldsymbol{\Phi}(i)}\right] \tag{5.103}$$

where $\boldsymbol{k}(i)$ is the identification gain vector, $\boldsymbol{P}(i)$ the identification (covariance) matrix, and α the forgetting factor.

To enhance the robustness of the estimation scheme as well as the speed of identification/adaptation, many improvements have been suggested in the literature; considering the formulae (5.101) – (5.103), some of the most promising ones are, for example, the constant trace version of the RLS estimation, [5], and the improved (robust) version of it introduced by Sripada and Fisher [25].

One important feature should be observed in these formulae: the regressor vector $\boldsymbol{\Phi}(i)$, the gain vector $\boldsymbol{k}(i)$ and the identification (covariance) matrix $\boldsymbol{P}(i)$ are the same for all p identification procedures for the p rows of the $\hat{\boldsymbol{\Theta}}$ parameter matrix. That means that in each identification step, the updating of $\boldsymbol{P}(i)$ and $\boldsymbol{k}(i)$ has to be performed only once applying (5.102) and (5.103). This version of $\boldsymbol{P}(i)$ and $\boldsymbol{k}(i)$ updating is sometimes called common regressors estimation.

In the controller-oriented model, the $\boldsymbol{\Psi}(i)$ vector is $\boldsymbol{\Psi}(i)$ of the model (5.85), and the regressor vector can be formed from the delayed and filtered $\boldsymbol{u}(i)$, $\boldsymbol{y}(i)$ and $\boldsymbol{w}^*(i)$, as well as from the vector of unity elements. The parameter matrix $\hat{\boldsymbol{\Theta}}$ can be formed from the polynomial matrices $\boldsymbol{R}_u$, $\boldsymbol{S}_u$ and $\boldsymbol{T}_u$, as well as the vector $\boldsymbol{q}$. In this case, identification should proceed as in the plant-oriented model case, i.e. using the common regressor estimation scheme.

However, it is obvious that the special structure of *unknown* and *known* factors of $\boldsymbol{R}$, $\boldsymbol{S}$ and $\boldsymbol{T}$ polynomial matrices can be utilized, as may be observed in formulae (5.86) and (5.87). Indeed, the number of identified parameters can be substantially reduced by removing zero elements from the *unknown* and *known* factors and performing the structure compression. However, by doing so, we encounter a situation in which the $\boldsymbol{\Phi}^j(i)$ regressor vectors are different for parameter estimation of the different rows of $\hat{\Theta}$. This specialized version of the multivariable RLS estimation scheme is called independent regressor estimation.

5.9 PERFORMANCE MEASURES FOR ADAPTIVE CONTROL ALGORITHMS

It seems reasonable to work out general and precise performance measures for adaptive control algorithms, which can be used to compare the results obtained by different adaptive schemes. Using such measures, we could answer the following typical questions:

- Does the new algorithm really offer much with respect to adaptive system performance when compared with existing ones?
- Which values of existing adaptive control schemes should we choose so as to achieve a stable, optimal adaptive system?

It is obvious that each stochastic adaptive control algorithm should ensure successful (optimum) control of a linear time-invariant stochastic plant, if possible. The concept of *maximum regulability*, defined below, can be used to gain performance-related information about the possibly adaptively controlled plant:

$$r_{max} = \frac{\left\{\begin{matrix}\text{Uncontrolled plant}\\ \text{output signal}\\ \text{variance}\end{matrix}\right\} - \left\{\begin{matrix}\text{Controlled signal}\\ \text{minimum}\\ \text{variance}\end{matrix}\right\}}{\left\{\begin{matrix}\text{Uncontrolled plant}\\ \text{output signal}\\ \text{variance}\end{matrix}\right\}}$$

Maximum regulability is a measure of the relative output variance decrease achievable in the best case, i.e. when a known plant (minimum phase or not) is controlled by the properly tuned minimum-variance controller. The maximum regulability indicates how much can be gained by controlling a stochastically disturbed, linear time-invariant plant. A small value of the maximum regulability index suggests that practically nothing can be gained by controlling that plant, because the small decrease of the output variance could well be below the noise variance of the measurement instruments.

The maximum regulability values for minimum phase (MP) plants can be computed using formulae first presented by Åström [2]:

$$r_{max} = \frac{\mathrm{E}\left\{\left[\frac{G}{A}e(i-k)\right]^2\right\}}{\mathrm{E}\left\{\left[\frac{C}{A}e(i-k)\right]^2\right\}}$$

An examination of the definition of r_{max} permits us to draw the following conclusions concerning its properties:

- r_{max} for minimum phase (MP) plants depends only on the time delay k and the properties of the disturbance filter $\frac{C}{A}$;

- r_{max} for NMP plants depends additionally upon the B^- polynomial, which could be justified on the basis of theoretical similarities between time delay and NMP;
- r_{max} for NMP plants is generally smaller than for their nearest MP neighbours, which could be concluded also from the similarities just mentioned;
- for white noise at the plant output we get $\frac{C}{A} = 1, G = 0$ and $r_{max} = 0$ which means a minimum of maximum regulability: the white-noise power spectrum is too broad for any control system to cope with;
- r_{max} is always smaller than 1, because in the best circumstances (i.e. for $k = 1$) the controlled variable is reduced to the driving white noise $e(i)$;
- r_{max} is invariant with respect to the white noise variance;
- with increasing time delay k, all other things being equal, r_{max} decreases to 0.

Another performance measure for adaptive control algorithms is the price paid for minimizing the output variance. This measure can be defined as the variance of the plant control signal for minimum-variance control and can be computed from

- for MP plants:

$$\mathrm{E}\left\{u^2(i)\right\} = \mathrm{E}\left\{\left[\frac{G}{B+}e(i)\right]^2\right\}$$

- for NMP plants:

$$\mathrm{E}\left\{u^2(i)\right\} = \mathrm{E}\left\{\left[\frac{G}{B+\tilde{B}-}e(i)\right]^2\right\}$$

It is also reasonable to define the relative minimum-variance control cost c_{MV} as:

$$c_{MV} = \frac{\left\{\begin{array}{c}\text{Control variance for}\\ \text{the minimum-variance}\\ \text{controlled plant}\end{array}\right\}}{\left\{\begin{array}{c}\text{Uncontrolled plant}\\ \text{output signal}\\ \text{variance}\end{array}\right\} - \left\{\begin{array}{c}\text{Controlled signal}\\ \text{minimum}\\ \text{variance}\end{array}\right\}}$$

thus relating the increase in the control signal variance to the corresponding decrease of the controlled signal variance.

All adaptive control algorithms are built on the basis of corresponding non-adaptive ones. It seems reasonable to define a performance measure with respect to perfectly calculated control algorithms of the chosen kind, i.e. assuming perfect knowledge of the plant structure and parameters as well as plant disturbance characteristics. Such a measure will be called the *obtainable regulability* r_o, and is defined as

$$r_o = \frac{\left\{\begin{array}{c}\text{Uncontrolled plant}\\ \text{output signal}\\ \text{variance}\end{array}\right\} - \left\{\begin{array}{c}\text{Controlled plant}\\ \text{output signal}\\ \text{variance}\end{array}\right\}}{\left\{\begin{array}{c}\text{Uncontrolled plant}\\ \text{output signal}\\ \text{variance}\end{array}\right\}}$$

It is obvious that:

$$r_o \leq r_{max}$$

Similarly it is worthwhile to account for control variable changes by introducing the *obtainable relative control cost*:

$$c_o = \frac{\left\{\begin{array}{c}\text{Control variance}\\ \text{for the controlled}\\ \text{plant}\end{array}\right\}}{\left\{\begin{array}{c}\text{Uncontrolled plant}\\ \text{output signal}\\ \text{variance}\end{array}\right\} - \left\{\begin{array}{c}\text{Controlled plant}\\ \text{output signal}\\ \text{variance}\end{array}\right\}}$$

These performance measures defined above may be evaluated for any control algorithm supposed to be implemented in the adaptive control system for the given plant.

The last concept concerns the truly adaptive control algorithm and determines its closeness to its non-adaptive origin. The *adaptive regulability* r_a could thus be defined as:

$$r_a = \frac{\left\{\begin{array}{c}\text{Uncontrolled plant}\\ \text{output signal}\\ \text{mean square}\end{array}\right\} - \left\{\begin{array}{c}\text{Adaptively controlled}\\ \text{plant output signal}\\ \text{mean square}\end{array}\right\}}{\left\{\begin{array}{c}\text{Uncontrolled plant}\\ \text{output signal}\\ \text{mean square}\end{array}\right\}}$$

Adaptive regulability should be determined in the stationary phase of the adaptive control system simulation, i.e. after the controller has converged. It is obvious that:

$$r_a \leq r_o.$$

Proceeding analogously we can define the relative obtainable control cost as a performance reference value for adaptive control based on a given non-adaptive control law. The *relative adaptive control cost* c_a could be defined as:

$$c_a = \frac{\left\{\begin{array}{c}\text{Mean square control}\\ \text{signal for the adaptively}\\ \text{controlled plant}\end{array}\right\}}{\left\{\begin{array}{c}\text{Uncontrolled plant}\\ \text{output signal}\\ \text{mean square}\end{array}\right\} - \left\{\begin{array}{c}\text{Adaptively controlled}\\ \text{plant output signal}\\ \text{mean square}\end{array}\right\}}$$

The r_a and c_a indices referred respectively to r_o and c_o give together an insight into how much are we behind the non-adaptive control performance in terms of quality and cost.

5.10 CONCLUSIONS

In this chapter several concepts and results concerning adaptive control systems were delivered. The field of stochastic adaptive control has gained a lot of interest among control research and development groups during the past few years. Several different adaptive control algorithms have emerged, many of which are heuristic, and almost all of which lacking proofs of control system stability and convergence of the parameter estimates. As one result of such rapid and difficult development, it turns out that only relatively simple adaptive controllers can be implemented in industry these days, as the more complicated ones are not robust enough. On the other hand, the huge degree of complication of the analysis and synthesis of adaptive control systems has stimulated the development of CAD-type software tools oriented to adaptive systems problems. Such software, built in the MATLAB/SIMULINK environment and considering the analysis and synthesis of minimum-variance and predictive adaptive control systems, will be presented in the following laboratory exercises.

5.11 LABORATORY EXERCISES

5.11.1 MATLAB software tools applied in laboratory course 5

No MATLAB toolboxes are directly used in the laboratory course 5. All examples provided are written as MATLAB m-files. SIMULINK is the basic tool. Controllers, which are elements of the simulated control systems, are prepared as S-functions. MATLAB allows creation of sophisticated software even though elements of this software are "hidden" in the SIMULINK structure. The S-functions realize all identification and synthesis of controllers. Only a few functions, from the Control Toolbox and Signal Processing Toolbox, are incorporated within S-functions.

5.11.2 Adaptive control

Minimum-variance control

Example 1 – the quality of control

The MV controller serves as an example of a self-tuner. The optimal value of the output signal variance is obtained in the steady state. Thus, it is important to evaluate the speed of convergence.

Use the MATLAB program `ad1_1` to check the speed of the output convergence. Calculate the exact value of the output variance and compare it with the value given in the final plot. The structure of the controller is chosen automatically, according to the chosen plant structure. The plant is assumed to be of ARX type.

Quick help

1. Enter program `ad1_1` at the MATLAB prompt.
2. According to the display in the command window, enter the parameters of the plant to be simulated as well as the parameters of the RLS algorithm and the duration of the simulation.
3. The program responds with the structure of the controller. The simulation runs and the output variance is plotted.
4. Repeat the simulation with different values of the RLS parameters.
5. Check whether the variance of the white noise used for disturbance generation affects the speed of convergence.

Questions

1. Does the variance of the white noise used to generate the disturbance affect the speed of convergence? Explain your answer.
2. How do the RLS parameters influence the speed of convergence?
3. How can you measure the speed of convergence, especially taking into account that the observed process is stochastic?

Example 2 – the influence of the structure choice on the speed of the output convergence

In the previous example the program chose the correct structure for the controller, i.e. proper orders of the controller polynomials. However, in reality we should expect differences between plant and model. Use program `ad1_2` to evaluate the influence of an improper choice of controller structure on the speed of convergence of the output variance.

Quick help

1. Enter program `ad1_2` at the MATLAB prompt.
2. Input the parameters of the plant to be simulated.
3. Enter the orders of the controller polynomials above or below the nominal values.
4. Set the RLS parameters: initial value of the diagonal of the covariance matrix 1000; forgetting factor 1; initial values of controller polynomials $R(z^{-1}) = 1, S(z^{-1}) = 1$; assumed dead time set equal to the plant dead time.
5. Run the experiment and repeat for different controller structures (orders of the controller polynomial).
6. Set the structure of the controller to be correct (according to the plant structure) and set the dead time greater or less than the plant dead time. Run the experiment for different assumed dead times.
7. In the all experiments, observe the plot of the output variance convergence.

Questions

1. How does wrong controller structure influence the speed of convergence of the output variance?

2. How does the steady-state error between the output variance and its minimal value depend on the wrong controller structure?

3. How does a wrong assumption about dead time influence the speed of convergence of the output variance?

Example 3 – the influence of the parameters of the identification procedure on the speed of the output variance convergence

In example 2, the parameters of the RLS method were kept constant. However, the behaviour of the self-tuners depends greatly upon the initialization (transient) phase in which the parameters of the controller are assumed to be completely unknown. Therefore it seems important to check the influence of the parameters of RLS on the performance in the sense of output variance convergence speed.

Quick help

1. Enter again program `ad1_2` at the MATLAB prompt.

2. Input the parameters of the plant to be simulated.

3. Enter the orders of the controller polynomials correctly according to the plant structure.

4. Set the initial parameters of the controller as $R(z^{-1}) = 1$, $S(z^{-1}) = 1$ and assume the dead time to be equal to the plant dead time.

5. Run the experiment for different values of the diagonal of the covariance matrix and forgetting factor.

6. In the all experiments, observe the plot of the output variance convergence.

Questions

1. What is the influence of the RLS parameters on the speed of convergence of the output variance?

2. Explain the influence of the forgetting factor (remember that the plant is time-invariant).

3. Try to propose a rule of thumb for the choice of RLS parameters.

Example 4 – non-minimum phase plant

One of the most important disadvantages of minimum-variance control is difficulty in control of NMP systems. Simple minimum-variance control leads to cancellation of the polynomial B and thus NMP zeros of the polynomial B give rise to unstable poles in the controlled systems. An *ad hoc* solution which provides a stable closed-loop system is to make the assumed dead time differ from the plant dead time. Another possibility is to use a weighted minimum-variance controller. Use program `adl_3` to check the first possibility and program `adl_4` to check the second.

Quick help

1. Enter program `adl_3` at the MATLAB prompt.
2. Input the parameters of the plant to be simulated.
3. Enter the orders of the controller polynomials correctly according to the plant structure.
4. Set the RLS parameters: initial value of the diagonal of the covariance matrix 1000; forgetting factor 1; initial values of the controller polynomials: $R(z^{-1}) = 1$, $S(z^{-1}) = 1$; assume the dead time is equal to the plant dead time.
5. Run the experiment and observe the performance of the system. Pay special attention to the control signal.
6. Run the experiment with different values of the assumed dead time (greater and less than the plant dead time).
7. Enter program `adl_4` at the MATLAB prompt.
8. Input the same parameters of the plant and RLS as in the previous case (assumed dead time should equal the plant dead time).
9. Run the experiment with different values of the weighting factor and observe the performance of the system.

Questions

1. Explain the reason for instability of the closed-loop system with the correct assumed dead time and simple minimum-variance control. In such a case, why is the instability first observed in the control action?
2. What value of assumed dead time seems best for stabilization of the system? Explain why shortening of the assumed dead time does not stabilize the system.
3. Explain the mechanism by which the weighting in the weighted minimum-variance control influences stability.
4. What is the influence of the weighting on the speed of convergence of the output variance?

5. Suggest the best choice of weighting factor.

Example 5 – time-varying plant

The previous examples concern time-invariant plants. This means that only the initial phase is important for parameter tuning. However, adaptive systems are designed mainly to cope with time-varying systems. The role of parameter tracking and thus of identification is crucial in such a system. Use program `adl_5` to check the quality of parameter tracking during plant changes.

Quick help

1. Enter program `adl_5` at the MATLAB prompt.
2. Input the parameters of the plant to be simulated. Note that a step change of the plant parameters is incorporated. Such a change, probably rare in practice, is convenient to observe parameter tracking.
3. Enter the orders of the controller polynomials correctly according to the plant structure.
4. Set the RLS parameters: initial value of the diagonal of the covariance matrix 1000; forgetting factor 1; initial values of controller polynomials $R(z^{-1}) = 1, S(z^{-1}) = 1$; assume the dead time is equal to the plant dead time.
5. Run the experiment and observe the convergence of the controller parameters.
6. Run the experiment with different values of the RLS parameters.

Questions

1. How do the parameters of RLS influence the convergence of the controller parameters?
2. Explain the slow parameter adaptation after the step plant parameter change, compared with initial adaptation.
3. Why does the disturbance, as the only excitation in the system, seem to be too weak for identification?

Tracking abilities of adaptive minimum-variance control

Minimum-variance control can follow the reference signal if the variance of the output error is minimized instead of that of the output signal itself. In the following, the tracking abilities of minimum-variance control will be investigated. The assumption is made that no disturbances affect the system.

Example 1 – reference signal tracking

If the reference signal (the only excitation in the system) does not change, the accuracy of identification deteriorates. Use program `ad2_1` to observe how the input signal influences the convergence of the model parameters.

Quick help

1. Enter program `ad2_1` at the MATLAB prompt.
2. Set the parameters of the plant to be simulated as well as the parameters of the RLS algorithm. Note that the plant is simulated as continuous-time; the sampling period and the offset should also be entered.
3. The initial values of the controller parameters should be $R(z^{-1}) = 1$ and $S(z^{-1}) = 1$.
4. Run the experiment and observe the signals being plotted. Note the quality of reference signal tracking.
5. Run the experiment again, changing the structure of the controller, assumed dead time and the RLS parameters. In every experiment, observe the quality of the reference signal tracking.
6. For the correct controller structure run a few experiments with different sampling periods. Observe the system behaviour between sampling.

Questions

1. How does the choice of controller structure affect the quality of reference signal tracking?
2. How do the RLS parameters influence tracking?
3. Explain the phenomenon that the continuous-time signal can be very large between sampling instants.

Example 2 – convergence of the parameters

A step-wise constant reference signal is useful to show the necessity of adequate excitation in identification. Use program `ad2_2` to check the convergence of the controller parameters.

Quick help

1. Enter program `ad2_2` at the MATLAB prompt.
2. Input the parameters of the plant to be simulated.
3. Enter the orders of the controller polynomials correctly according to the plant structure.

4. Set the RLS parameters: initial value of the diagonal of the covariance matrix 1000; forgetting factor 1; initial values of controller polynomials $R(z^{-1}) = 1, S(z^{-1}) = 1$; assume the dead time is equal to the plant dead time.

5. Run the experiment and observe the convergence of the controller parameters.

6. Run the experiment with different value of the RLS parameters.

7. Repeat the experiments but after the controller parameters converge change the plant parameters then continue the experiment.

8. Repeat the experiment with over- and underparameterized controller structures as well as with incorrect assumed dead times. In every experiment observe the convergence of the controller parameters.

Questions

1. How do the parameters of RLS influence the convergence of the controller parameters?

2. Explain the slow parameter adaptation after the step plant parameter change, compared with initial adaptation. Comment on the role of the forgetting factor.

3. Compare the results observed with those obtained in example 5 of the previous section (time-varying systems).

Example 3 – the influence of weighting

Weighting of the control signal improves robustness of the system with respect to stability. However, it causes the performance of the system to deteriorate. Use program `ad2_3` to check the influence of control signal weighting on control quality.

Quick help

1. Enter program `ad2_3` at the MATLAB prompt.

2. Input the parameters of the plant to be simulated.

3. Enter the orders of the controller polynomials correctly according to the plant structure.

4. Set the RLS parameters: initial value of the diagonal of the covariance matrix 1000; forgetting factor 1; initial values of controller polynomials $R(z^{-1}) = 1, S(z^{-1}) = 1$; assume the dead time is equal to the plant dead time.

5. Run the experiment for different values of the weighting factor and observe the plots of reference, input and output signals.

Questions

1. How does the weighting factor affect the system performance?

2. Suggest how we should choose the weighting factor with respect to the system performance.

Adaptive GPC – tracking properties

Adaptive generalized predictive control (GPC) serves as an example of indirect adaptive control. Thus, the common difference from the previously discussed minimum-variance control is the necessity of identification of the plant model and recalculation of the controller parameters. GPC provides integral action. However, the quality of the reference signal tracking depends heavily on the parameters of GPC.

Example 1 – correct structure of the model

Use MATLAB program `ad3_1` to check the quality of the reference signal tracking. The main purpose of the example is to show that different control and output prediction horizons can provide stable behaviour of the system.

Quick help

1. Enter program `ad3_1` at the MATLAB prompt.

2. Set the parameters of the plant to be simulated. Choose the plant to be minimum phase.

3. Set the parameters of the controller as $N_1 = 1$, $N_2 = 1$, $N_u = 1$, no control weighting and no feed-forward precompensation.

4. Set the initial values of the model parameters as $B(z^{-1}) = 1$, $A(z^{-1}) = 1$.

5. Set the RLS parameters: initial value of the diagonal of the covariance matrix 1000; forgetting factor 1.

6. Run the program and observe the performance of the system.

7. Repeat the experiment changing the horizons N_1, N_2 and N_u. Try to find the best set of horizons.

8. Change the plant to be NMP and observe the non-stable behaviour of the system for$N_1 = 1$, $N_2 = 1$, $N_u = 1$. Try to stabilize the system by extending horizon N_2 and then make the performance better by extending horizon N_u.

9. Repeat the previous point with $N_1 = 1$, $N_2 = N_u$ and with control signal weighting. Try to find a weighting factor providing stable performance of the system.

10. Repeat the experiments with feed-forward precompensation.

Questions

1. Explain why GPC behaves like a minimum-variance controller providing that $N_2 = N_u$ and no control weight is incorporated.
2. Why cannot a feed-forward precompensator change stability conditions? Explain the role of the feed-forward precompensator. Compare the effect of precompensation with control signal weighting.
3. Why can setting $N_2 > N_u$ stabilize the system with a nonminimum phase plant?

Example 2 – influence of tuning parameters

Use MATLAB program `ad3_1` to check the quality of the reference signal tracking for different GPC parameters.

Quick help

1. Enter program `ad3_1` at the MATLAB prompt.
2. Set the parameters of the plant to be simulated. Choose the plant the same as in the previous example (minimum or non-minimum phase).
3. Use the best set of horizons according to the previous example.
4. Set initial values of model parameters as $B(z^{-1}) = 1$, $A(z^{-1}) = 1$.
5. Set the RLS parameters: initial value of the diagonal of the covariance matrix 1000; forgetting factor 1.
6. Run the program, observe the performance of the system and repeat the experiment, changing the parameters of the controller, namely feed-forward speed precompensator and control weighting parameter.

Questions

1. Why does slowing the response speed of the precompensation lead to system behaviour similar to that obtained by increasing the value of the control weighting parameter?
2. Are the two methods of controller tuning the same? Why?
3. Which of tuning method do you suggest implementing in the real world? Explain.

Example 3 – incorrect model structure

Use MATLAB program `ad3_1` to check the quality of the reference signal tracking assuming the wrong model structure.

Quick help

1. Enter program ad3_1 at the MATLAB prompt.

2. Set the parameters of the plant to be simulated. Choose the plant to be minimum phase. For comparison, choose the same plant as in the previous example.

3. Set the parameters of the controller to the best values you found in the previous example.

4. Set the RLS parameters: initial value of the diagonal of the covariance matrix 1000; forgetting factor 1.

5. Run the program for different model structures and observe the performance of the system.

Questions

1. How do under- and overparameterization of the model influence the quality of reference signal tracking?

2. Explain why too small an order of the model's B polynomial can cause instability, and how we can avoid this effect without using the larger order for the B polynomial.

3. Why cannot the reference signal tracking be perfect even when the structure of the model is perfect?

Disturbance rejection in adaptive predictive control

Again, GPC, an example of an adaptive predictive controller, will be used. The following examples shows the ability to reject both deterministic and stochastic disturbances.

Example 1 – rejection of bias disturbance

A constant disturbance affecting the output of the plant to be controlled is one of most typical in practice. The disturbance is often called a bias. Use MATLAB program ad3_2 to check the quality of the reference signal tracking if bias acts.

Quick help

1. Enter program ad3_2 at the MATLAB prompt.

2. Set the parameters of the plant, controller and RLS algorithm.

3. Run the program assuming no bias disturbance and observe the quality of the reference signal tracking.

4. Repeat the experiment assuming bias. Observe the deterioration in quality of the reference signal tracking.

5. Repeat the experiment assuming that bias begins after the model parameters converge. Compare the deterioration of the reference signal tracking with the previous experiment.

Questions

1. Explain why the greatest deterioration of the reference signal tracking can be observed when the bias begins after the model parameters converge.

2. Discuss the following possibilities for improving the system performance: identification of the controller's free term, reinitialization of the RLS covariance matrix, keeping the trace of the RLS covariance matrix constant.

3. Try to implement one of these solutions in program `ad3_2` and compare the reference signal tracking with the results obtained previously.

Example 2 – rejection of stochastic disturbance

Generalization of minimum-variance control in the sense of the GPC algorithm causes the quality of stochastic disturbance rejection to deteriorate compared with MV. However, it is well known that proper performance of MV needs an almost perfect model, while predictive control is much more robust. Anyway, it is worth checking how predictive control can cope with stochastic disturbances. Use MATLAB program `ad3_3` to check how the quality of the rejection of stochastic disturbances deteriorates when compared with minimum-variance control.

Quick help

1. Enter program `ad3_3` at the MATLAB prompt.

2. Set the parameters of the plant, MV controller, GPC controller and RLS algorithm. Remember that, in the case of NMP plant, a weighted minimum-variance algorithm should be used. This leads to the necessity of choosing the weighting parameter. The parameter influences the quality of disturbance rejection.

3. Run the program and observe the quality of stochastic disturbance rejection in the MV and GPC cases. Pay special attention pay to the final results of the experiment, examining the values of output variances.

4. Repeat the experiment, assuming different values for the GPC parameters.

Questions

1. Explain the deterioration of the stochastic disturbance rejection in the GPC case.

2. How do the parameters of the GPC controller influence the stochastic disturbance rejection performance.

3. Compare the quality of rejection for the same value of control weighting parameter (for MV and GPC controllers).

Non-stationary plant

An adaptive predictive controller can cope well with time-varying systems. It is interesting to test the convergence of the model parameters and the algorithm's ability to track the plant parameters.

Example 1 – convergence of model parameters

Use MATLAB program `ad3_4` to check the convergence of the model parameters.

Quick help

1. Enter program `ad3_4` at the MATLAB prompt.
2. Set the parameters of the plant, controller and RLS algorithm.
3. Run the program assuming correct model structure. Observe the convergence of the model parameters.
4. Repeat the experiment, assuming different parameters in the controller and RLS algorithm.

Questions

1. How does the convergence of the model parameters depend on the parameters of the controller. Which of the controller parameters is the crucial one?
2. Explain the influence of the initial value of the covariance matrix of the RLS algorithm on convergence.

Example 2 – convergence of model parameters after plant change

Use MATLAB program `ad3_4` to check the convergence of model parameters after a plant change. Program `ad3_5` allows also testing of parameter tracking, when the change is slow.

Quick help

1. Enter program `ad3_4` at the MATLAB prompt.
2. Set the parameters of the plant, controller and RLS algorithm.
3. Run the program assuming correct model structure. Observe the convergence of the model parameters. Once the model parameters are more or less constant, change one of the plant parameters (e.g. the gain). Observe the convergence after the change.
4. Repeat the experiment assuming different parameters of the controller and RLS algorithm.
5. Repeat the experiment assuming a slow plant parameter change.

Questions

1. Explain why the convergence of the model parameters does not depend significantly on the controller parameters.

2. Explain why the initial value of the covariance matrix of the RLS algorithm has little influence on the convergence after a plant change.

3. Discuss the possibility of improved convergence after a plant change.

5.12 REFERENCES

[1] A. Albert and R. Sittler. A method for computing least squares estimators that keep up with the data. *SIAM J. Control*, **3**(3), 384–417, 1966.

[2] K. J. Åström. *Introduction to Stochastic Control Theory*. Academic Press, New York, 1970.

[3] K. J. Åström. Simple self-tuners. Lund Report CODEN: LUTF2/(TRFT-7184), Lund Institute, Lund, Sweden, 1979.

[4] K. J. Åström and B. Wittenmark. *Computer Controlled Systems – Theory and Design.* Prentice Hall, Englewood Cliffs, NJ, 1984.

[5] K. J. Åström and B. Wittenmark. *Adaptive Control*. Addison-Wesley, Reading, MA, 1989.

[6] U. Borisson. Self-tuning regulators for a class of multivariable systems. *Automatica*, **15**(2), 209–215, 1979.

[7] C. R. Cutler and B. C. Ramaker. Dynamic matrix control – a computer control algorithm. In *Joint Automatic Control Conference*, paper WP5–B, San Francisco, CA, 1980.

[8] L. Dugard and J. M. Dion. Direct adaptive control for linear multivariable systems. *Int. J. Control*, **42**, 1251–1281, 1985.

[9] L. Dugard, G. C. Goodwin, and X. Xianya. The role of the interactor matrix in multivariable stochastic adaptive control. *Automatica*, **20**, 701–709, 1984.

[10] T. R. Fortescue, L. S. Kershenbaum, and B. E. Ydstie. Implementation of self-tuning regulators with variable forgetting factors. *Automatica*, **17**, 831–835, 1981.

[11] G. C. Goodwin, P. J. Ramadge, and P. E. Caines. Discrete-time multivariable adaptive control. *IEEE Trans. Automat. Contr.*, **AC-25**, 449–455, 1980.

[12] G. C. Goodwin and K. S. Sin. *Adaptive Filtering, Prediction and Control.* Prentice Hall, Englewood Cliffs, NJ, 1984.

[13] X. Y. Gu, W. Wang, and M. M. Liu. Multivariable generalized predictive adaptive control. In *Proceedings of the 1991 American Control Conference*, pp. 1746–1751, Boston, MA, 1991.

[14] R. Isermann, K. H. Lachmann, and D. Matko. *Adaptive Control Systems.* Prentice Hall, Hemel Hempstead, 1992.

[15] L. S. Kershenbaum and T. R. Fortescue. Implementation of on-line control in chemical process plants. *Automatica*, **17**(6), 777–788, 1981.

[16] M. Kinnaert. Generalized predictive control of multivariable linear systems. In *Proceedings of the 26th Conference on Decision and Control*, pp. 1247–1248, Los Angeles, CA, 1987.

[17] H. N. Koivo. A multivariable self-tuning controller. *Automatica*, **16**, 351–366, 1980.

[18] C. L. Lawson and R. J. Hanson. *Solving Least Squares Problems.* Prentice Hall, Englewood Cliffs, NJ, 1974.

[19] L. Ljung and T. Söderström. *Theory and Practice of Recursive Identification.* MIT Press, Cambridge, MA, 1983.

[20] K. S. Narendra and A. M. Annaswamy. *Stable Adaptive Systems.* Prentice Hall, Englewood Cliffs, NJ, 1989.

[21] A. Niederliński and J. Mościński. Robust implicit adaptive control for nonminimum phase plants. In *Proceedings of the IFAC Workshop on Robust Adaptive Control*, pp. 163–169, Newcastle, Australia, 1988.

[22] R. Scattolini. Multirate self-tuning control of multivariable systems. In *Preprints of the IFAC Symposium on Adaptive Systems in Control and Signal Processing*, vol. II, pp. 365–370, Glasgow, 1989.

[23] R. Scattolini and N. Schiavoni. Generalized minimum variance control of MIMO systems – a stability result. *Automatica*, **23**(6), 797–799, 1987.

[24] S. L. Shah, C. Mohtadi, and D. W. Clarke. Multivariable adaptive control without prior knowledge of the delay matrix. *Syst. Control Lett.*, **9**, 295–306, 1987.

[25] N. R. Sripada and D. G. Fisher. Improved least squares identification. *Int. J. Control*, **46**, 1889–1913, 1987.

[26] P. E. Wellstead, and M. B. Zarrop. *Self-Tuning Systems. Control and Signal Processing.* Wiley, Chichester, 1991.